Diamond
Katana DA20
A Pilot's Guide

Ed Helmick

Includes information on
DA20-A1 and DA20-C1 Models

Aviation Supplies & Academics, Inc.
Newcastle, Washington

Diamond Katana DA20: A Pilot's Guide
by Ed Helmick
© 1998 Aviation Supplies & Academics, Inc.

Diamond Katana photography used with permission of Diamond Aircraft Industries.

All rights reserved. No part of this publication may be reproduced, stored in a retrieval system, or transmitted in any form or by any means, electronic, mechanical, photocopy, recording or otherwise, without the prior written permission of the copyright holder. While every precaution has been taken in the preparation of this book, the publisher and Ed Helmick assume no responsibility for errors or omissions. Neither is any liability assumed for damages resulting from the use of the information contained herein.

Aviation Supplies & Academics, Inc.
Newcastle, Washington

Printed in the United States of America

01 00 99 98   9 8 7 6 5 4 3 2 1

ASA-PG-DA20
ISBN 1-56027-321-6

Helmick, Ed.
    Diamond Katana DA20 / Ed Helmick.
        p.   cm. — (A pilot's guide.)
    "Includes information on Diamond Katana DA20-A1 and DA20-C1 models."
    Includes index.
    ISBN 1-56027-321-6
    1. Katana DA20 (Training plane) 2. Private flying.
    I. Title. II. Title: Diamond Katana DA twenty. III. Series.
    TL686.D53H45 1998
    629.132'5243—dc21   98-51885
                                           CIP

Contents

Editor's Note vi

Section 1 – General Description
Introduction to the Diamond Katana 1-3
The Airframe 1-6
The Flight Controls 1-8
The Landing Gear 1-10
The Engine 1-11
The Propeller 1-16
The Ignition System 1-18
The Oil System 1-19
The Starter System 1-20
The Fuel System 1-21
The Carburetors 1-24
The Electrical System 1-26
The Stall Warning System 1-30
The Lighting System 1-31
The Avionics 1-32
The Gyroscopic Instruments 1-33
The Pitot-Static System 1-34
The Heating and Ventilation System 1-35
Seats, Harnesses and Baggage Compartment 1-36
The Canopy 1-37

Section 2 – Limitations
Diamond Katana DA20-A1 Dimensions 2-3
Diamond Katana DA20-C1 Dimensions 2-5
The "V" Airspeed Code 2-6
Diamond Katana DA20-A1 Limitations 2-6
    Airspeed Limitations 2-6
    Airspeed Indicator Markings 2-6
    Maximum Demonstrated Crosswind Component 2-6
    Airframe Limitations 2-6
Diamond Katana DA20-C1 Limitations 2-7
    Airspeed Limitations 2-7
    Airspeed Indicator Markings 2-7
    Maximum Demonstrated Crosswind Component 2-7
    Airframe Limitations 2-7

DA20-A1 and DA20-C1 2-8
    Flight Load Factors 2-8
    Performance Limitations 2-8
    Kinds of Operation 2-8
    Minimum Equipment List 2-8
DA20-A1 2-9
    Engine Limitations 2-9
    Oil Capacity 2-10
    Oil Grade 2-10
    Coolant Type 2-10
    Fuel Grade 2-10
    Miscellaneous Limitations 2-10
DA20-C1 2-11
    Engine Limitations 2-11
    Oil Capacity 2-12
    Oil Grade 2-12
    Fuel Grade 2-12
    Miscellaneous Limitations 2-12

Section 3 – Handling the Katana DA20

    Ground Handling 3-3
    Entering the Katana 3-4
    Engine Starting 3-4
    Taxiing 3-7
    Power and Pre-Takeoff Checks 3-8
    Takeoff 3-9
    Climbing 3-10
    Cruise Flight 3-10
    Engine Handling 3-10
    Aerodynamic Stalls 3-11
    Spins 3-13
    Descent 3-14
    Landing 3-15
    After Landing 3-16
    Parking and Tie Down 3-16

Section 4 – Mixture and Carburetor Icing Supplement

    Introduction 4-3
    Carburetor Icing 4-3
    How Carburetor Icing Forms 4-3
    Conditions Likely to Lead to Carburetor Icing 4-4
    Carburetor Icing Conditions 4-5
    Symptoms of Carburetor Icing 4-6
    Use of Carburetor Heat 4-6

Section 5 – Katana DA20 Checklists

Approaching Aircraft 5-3
Cockpit 5-4
Left Main Landing Gear 5-5
Left Wing 5-5
Fuselage 5-5
Empennage 5-5
Right Wing 5-6
Right Main Landing Gear 5-6
Nose 5-6
Before Starting Engine 5-7
Starting Engine 5-8
Before Taxiing 5-9
Taxiing 5-9
Before Takeoff 5-9
Takeoff 5-11
    Normal Takeoff 5-11
    Short-Field Takeoff 5-12
    Soft-Field Takeoff 5-12
Climb 5-12
Cruise 5-13
Descent 5-13
Accelerated/Fast Descent 5-13
Landing Approach 5-13
    Soft-Field Landing 5-14
    Short-Field Landing 5-14
    Forward Slips to a Landing 5-14
    Balked Landing/Go Around 5-15
After Landing and Clearing the Runway 5-15
Engine Shut-down 5-15

Section 6 – Katana DA20 Loading and Performance

Loading 6-3
Mathematical Weight and Balance Calculation 6-5
Performance 6-6
Takeoff and Landing Performance Graphs 6-9
Enroute Performance Graphs 6-10
Runway Dimensions 6-10
    Takeoff Run Available (TORA) 6-10
    Accelerate/Stop Distance (A/SD) 6-10
    Takeoff Distance Available (TODA) 6-10
    Landing Distance Available (LDA) 6-10

Section 7 – Conversions

Takeoff Distance Factors 7-3
Landing Distance Factors 7-4
Runway Contamination 7-5
Use of the Wind Component Graph 7-6
Wind Component Graph 7-7
Temperature 7-8
Distance – Meters/Feet 7-9
Distance – Nautical Miles/Statute Miles 7-10
Volume (Fluid) 7-11

Index

Editor's Note

Welcome to ASA's *A Pilot's Guide* series. In this guide, you'll learn from the experts the general principles involved in flying the Diamond Katana DA20, with extra insight on individual characteristics gleaned from flying experience. This Katana DA20 guide covers both models—the DA20-A1, and the DA20-C1. Where the features differ between models, the text is clearly marked to show individual characteristics.

Diamond Katana DA20: A Pilot's Guide is not an authoritative document. Material in this book is presented for the purposes of orientation, familiarization, and comparison only.

Performance figures are based upon the indicated weights, standard atmospheric conditions, level hard-surface dry runways, and no wind. They are values based upon calculations derived from flight tests conducted by the aircraft manufacturer under carefully documented conditions and using professional test pilots. Performance will vary with individual aircraft and numerous other factors affecting flight.

The approved *Pilot's Operating Handbook* or the approved *Airplane Flight Manual* is the only source of authoritative information for any individual aircraft. In the interests of safety and good airmanship, the pilot should be familiar with these documents.

About the Author

Ed Helmick, a flight instructor since 1985, flew the first DV20 Katana prototypes brought to North America in the spring of 1993. Hired by Diamond Aircraft Corporation to demonstrate the Katana, Ed accumulated 1,700 hours flying the DV/DA20 and subsequently opened two flight schools based on the airplane. Ed has more experience with the Katana than anyone; moreover, through the years he has developed an affinity for this aircraft that highly complements his knowledge of its qualities and idiosyncrasies, detailed in his *Diamond Katana DA20* volume of *A Pilot's Guide* Series.

Section 1
General Description

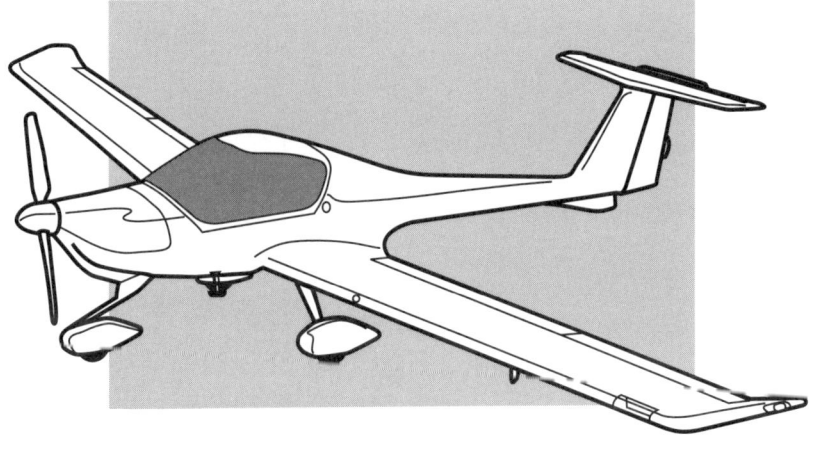

Diamond Katana DA20: A Pilot's Guide

Section 1 **General Description**

Introduction to the Diamond Katana

The Diamond Katana has become a very popular training and light touring aircraft. At the time of this writing there are about 400 Katanas operating throughout the world. The Katana is the result of the vision and financial commitment of a German industrialist and automotive distributor with a passion for aviation. He saw an exciting future for airplanes of a modern composite material design powered by modern engines. The Katana was conceived as the first of a family of new aircraft, and is proving that there is appeal and a market for a new generation of aircraft.

The story of the Katana begins in 1978 with the Hoffmann Flugzeugbau GmbH company in Dachau, Germany, a company formed to design and build a new motorglider, designated the H.36. In 1980, after the prototype had been test flown, a partnership was developed with plans to build the H.36 in the Israeli town of Dimona. These plans never materialized, but the name "Dimona" stuck. In 1981 production of the Dimona was moved to Carinthia, in southern Austria. By 1984, over 200 Dimona motorgliders were built. The company was reorganized as Hoffmann Aircraft Limited and later acquired by the state-run Austrian railway company.

In 1990 a totally new version of the motorglider was developed and designated the HK.36 Super Dimona. It was powered by the then new Bombardier Rotax 912 aircraft engine. At this time the state wished to divest itself of the aviation business, which was then acquired as a private investment to become a major airframe manufacturer and renamed HOAC Austria. One of the company's first actions was to develop a short-winged version of the Super Dimona motorglider. Designated the LF.2000, this aircraft was the proof-of-concept model for the two-seat trainer and light-touring airplane that was to become the Katana. The tricycle landing-gear version of the airplane first flew in December of 1991.

When a tricycle undercarriage was added to the LF.2000 design, it was renamed the DV.20 Katana. (The "D" and the "V" came from the name of the new company owner and his associate.) The "20" is shortened from the "2000" used on the conventional landing-gear prototype. The name Katana is the Japanese word for the smaller of the Samurai swords and is pronounced, kuh-TAN-ah. The Canadian and United States "Type Certificate" for the DV20 Katana was issued on March 1, 1994.

Diamond Katana DA20: A Pilot's Guide

PRODUCTION YEAR	MODEL	MODEL NAME
1994 – Present	DA20-A1	Katana
1997 – Present	DA20-C1	Katana

Realizing that the greatest market potential for the aircraft was in Canada and the United States, the company began searching for a North American manufacturing site. In February of 1992, London, Ontario, Canada was selected as the location for a 256,000-square foot modern aircraft production facility. Between July 1993 and July 1994, the transfer of technology for composite material aircraft manufacturing was completed for the production of the Canadian- built Katana. This Katana DA20-A1 received its Canadian CAA Type Certificate in July of 1994 and its U.S. FAA Type Certificate in December of 1994.

Bringing the airplane to production in Canada allowed for the incorporation of over 40 refinements and improvements to the Austrian version of the Katana, and the designation DA20-A1. The official maiden flight of the DA20 Katana was on June 29, 1994. The DA originally stood for Dimona Aircraft, until the company name was changed to Diamond Aircraft in June of 1994. Sole production facility for the Katana is now located at London, Ontario, Canada where the aircraft are manufactured on an assembly-line basis to meet worldwide demand.

Section 1 **General Description**

The official designation of the first Canadian-built Katana was Airplane Model DA20-A1 with a Canadian Type Certificate number of A-191, and a U.S. Type Certificate number of TA4CH. In working with Air Traffic Control (ATC) there has been some confusion with the DA20 Falcon executive jet, with questions asked about the 80-knot airport traffic pattern speed. It has been suggested by some in ATC that all Katana pilots use the DV20 designation to avoid confusion in their system. The International Civil Aviation Organization (ICAO) has changed the designation for flight planning purposes for all Katanas to DV20, to help reduce this confusion.

This book deals only with the Canadian production DA20 Katana that is now being delivered to flight schools large and small throughout the world. The DA20-C1 Type Certificate for both Canada and the U.S. was issued in April of 1998 and amended for an additional propeller option in September of 1998. As this book goes to print, we have learned that Diamond Aircraft intends to make the Sensenich climb propeller standard equipment on all DA 20-C1 aircraft after serial number 35.

Diamond Katana DA20: A Pilot's Guide

The Airframe

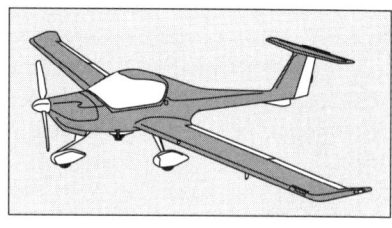

The DA20 Katana is an all-composite construction airplane with the distinguishing features being the low wing, "T" tail design with a large bubble canopy for unsurpassed cockpit visibility. The primary construction and strength of the airframe is a result of combining plastic resins with high-strength lightweight filaments. In the Katana, most of the structure is fiberglass with carbon fiber in high-stress areas. This makes for a very strong and durable airframe that can be molded into sleek and glassy-smooth aerodynamic designs for reduced drag.

The parent company has a long history of building strong and efficient airframes for motorgliders using composite material technology. There is very little metal in the Katana airframe and it is interesting to note that the Katana is one of the most highly produced, factory certified, general aviation production airplanes in the world whose primary structure is based on composite material technology. The use of composite materials allows for a shape that is modern and distinctive.

The fuselage, including the vertical stabilizer and the ventral fin, is assembled as two pre-molded and oven-cured glass fiber reinforced plastic (GFRP) shells. These shells are bonded together in a semi-monocoque construction with bulkheads and stiffeners. The function of the ventral fin is to prevent the rudder from striking the ground in a tail low landing. A replaceable Teflon skid is provided at the bottom of the ventral fin. The forward firewall bulkhead is constructed of glass-fiber and carbon-fiber reinforced plastic, with a special fire retardant fleece covered by a stainless steel plate. The length of the Katana is 23.9 feet.

The cantilever wing is constructed, similar to the fuselage, with a conventional arrangement of ribs along a spar. The wing incorporates an "I" beam-shaped GFRP spar with a carbon fiber cap on the top and bottom. The wing spar is attached using three bolts, and reaches to the middle of the fuselage through the spar box. The turned-up wing tips, called "vortips," improve roll stability, reduce drag and enhance aileron effectiveness at low airspeed.

Section 1 **General Description**

The Wortmann FX 63-137/20 HOAC airfoil was developed in conjunction with the University of Stuttgart (Germany) as a high-lift, low-drag laminar flow wing with excellent low-speed handling qualities. The Katana has a wingspan of 35.6 feet including the wing-tip strobe (ACL) lights. Because of the narrow, high aspect ratio wing and slender, low-drag fuselage, many people comment about the Katana having a long wing. The Katana wing is actually shorter than the wingspan of a Cessna 172. The Katana wing has a dihedral of 4 degrees, and the leading edge is swept back one degree.

The horizontal stabilizer is also of semimonocoque construction with ribs and a spar. A locating pin and four bolts secure the horizontal stabilizer to the top of the vertical stabilizer. By design, the Katana "T" tail provides excellent low-speed pitch authority while minimizing the effects of propeller wind blast or slipstream. The ailerons, flaps, and elevator are all of carbon fiber construction. The entire airframe is painted with a high-quality primer and white Acrylic Enamel paint to protect it against moisture and ultraviolet rays. There is no life limit on the airframe; however, the FAA Type Certificate requires a comprehensive airframe inspection after 6,000 hours of flight time to ensure airworthiness for further operation.

The airframe certification did include limits for outside air temperature for operation of the airplane and a maximum structural temperature. The minimum temperature the airplane can be flown in is -31°F (-35°C) outside air temperature. The maximum structural temperature is 131°F (55°C). A structural temperature indicator is located on the spar bridge and is accessed by lifting the flap between the seat back cushions. The indicator is visible through the cut-out in the seat shell backs, and only requires checking when the outside air temperature exceeds 100°F (38°C).

"Vortip" design

The Flight Controls

Push/pull rods deflect the ailerons and elevator and provide excellent control responsiveness and low maintenance. Under the wing, Plexiglas inspection panels are provided to view the control surface bellcranks. The ailerons deflect up 16 degrees and down 13 degrees. The deflection of the elevator is up 16 degrees and down 14 degrees. Only on the rudder are control cables used with a deflection of 30 degrees left and right.

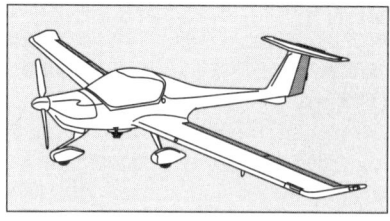

The seating is fixed and the rudder pedal distance adjusts for different height people, which is unique to the Katana among general aviation aircraft. The wing flaps are electrically actuated with a three-position switch located on the lower center of the instrument panel. The three flap positions are cruise (0 degrees to the trailing edge), take-off (15 degrees down) and landing (40 degrees down on A1 model and 45 degrees down on C1 model). Pitch forces may be trimmed out using a rocker switch

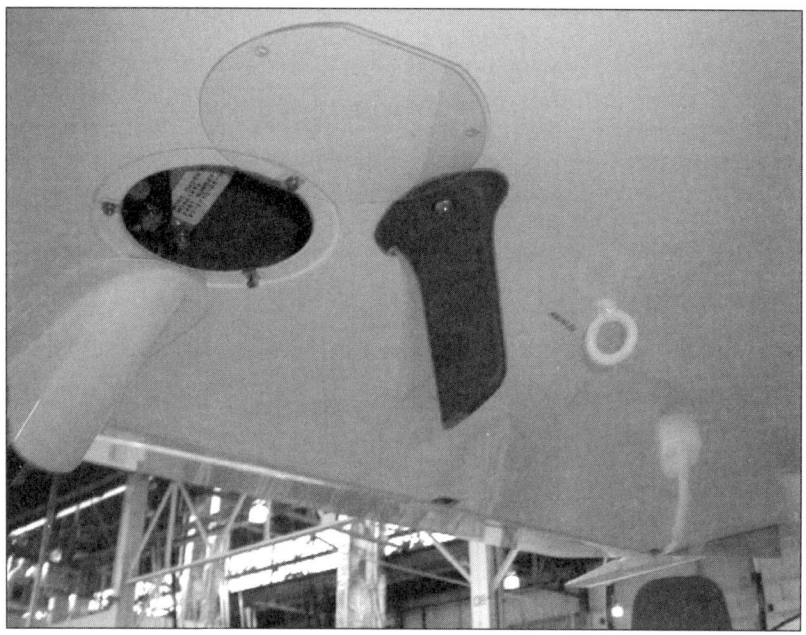

The aileron bellcrank inspection panel, under the wing

Section 1 **General Description**

located behind the throttle quadrant which electrically activates an anti-servo tab on the elevator of the A1 model and applies a load to compression springs of the elevator push rod on the C1 model. The trim indicator is a vertical light bar located at the top center of the instrument panel.

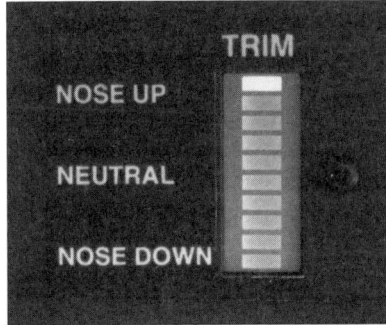

Trim indicator

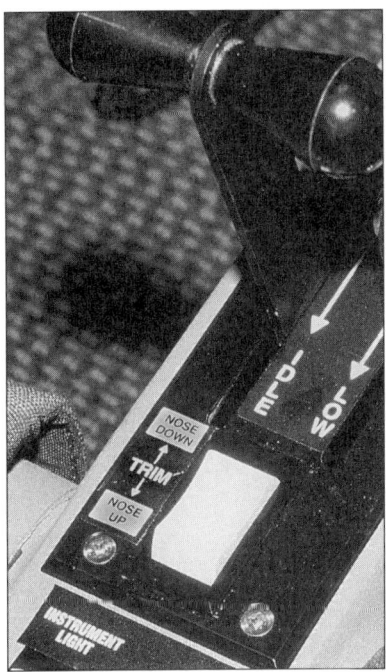

Trim rocker switch on throttle quadrant

Flaps switch

1-9

The Landing Gear

The main landing gear is either spring-steel or spring aluminum, depending on the serial, and are attached to the fuselage spar box with four bolts each. The main wheels are equipped with hydraulically operated disk brakes. Tire size for the main gear is 15 X 6.00-5. The nose tire was 4.00-4 for the first 50 aircraft produced. This increased to a more common 5.00-4 on serial number 10051 and subsequent aircraft. The nose-gear assembly is attached to the engine mount and the bottom of the fuselage. An elastomer package (these look like rubber donuts of varying sizes) located between the engine mount and the steel gear leg provides nose gear shock absorbing. Normal tire pressure is 33 psi for the main tires and 26 psi for the nose tire.

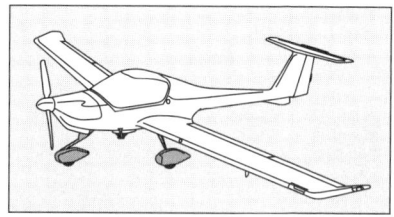

Main landing gear

The main gear has a track of 6 feet 2.8 inches and the wheel base between the nose tire and the main tires is 5 feet 8.9 inches. Steering the Katana on the ground is accomplished by use of differential brake pressure with a nose wheel that will caster 30 degrees left and right of center. Steering friction prevents nose wheel shimmy.

Nose gear

The parking brake is a small knob located on the lower panel directly forward of the throttle quadrant. Applying pressure to the toe-brake pedals and pulling the knob to the out position sets the parking brake. This closes a valve and locks brake pressure in the hydraulic brake lines. To release the parking brake, the knob is pushed in toward the instrument panel.

Section 1 **General Description**

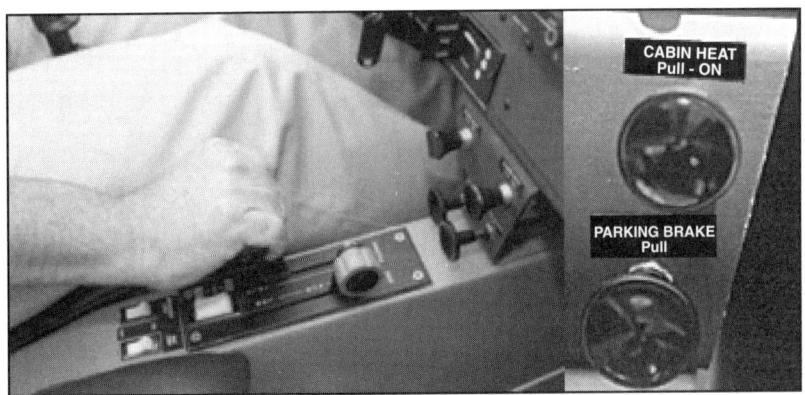

Parking brake knob on panel directly forward of throttle quadrant

The Engine
DA20-A1

The powerplant for the DA20-A1 Katana is the modern Bombardier Rotax 912 F3 engine that develops 81 hp. The 4-cylinder, 73.9 cubic inches (1.211 liters), 4-stroke, horizontally-opposed aircraft engine became fully certified in the United States under 14 CFR Part 33 in February of 1995. The engine uses a dry-sump forced-lubrication system. Typical of aircraft engines, each cylinder has two spark plugs; however, unique to the Rotax 912, a low-maintenance dual electronic capacitor discharge ignition system is used on this modern airplane. Some early serial number Canadian DA20s and all Austrian-built DV20 Katanas were powered by the A3 version of the Rotax 912. At the time the 912 F3 engine was certified, the FAA issued a letter recognizing the A3 engine as equivalent because the common engine core provides the same level of safety.

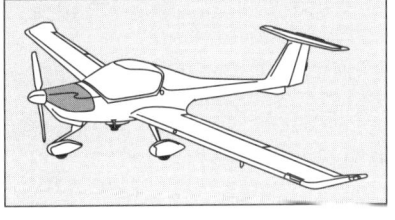

The Bombardier Rotax 912, specifically designed as an aircraft engine in 1990, brings to general aviation some interesting features. Most notable is that it is water-cooled, and you will notice a small radiator at the front of the cowling (*see* photograph on next page). Actually, semi-water-cooled—the engine uses a combination of water, air and oil for cooling. The cylinder heads are cooled by a water jacket, while the cylinders themselves are finned and cooled by ram air in the traditional manner with a small, round air-intake duct on the front of the cowling. The engine also has an oil cooler to contribute to normal operating temperatures, with an

1-11

Diamond Katana DA20: A Pilot's Guide

Rotax 912 engine installed in DA20-A1

Section 1 **General Description**

intake vent on the lower right cowling. The advantage of water-cooling the cylinder heads is that it eliminates shock-cooling to the valves, by providing a more stable temperature for longer life and lower maintenance. The water jacket around the valves also contributes to lower engine noise levels.

The engine is a high-rpm engine which on takeoff turns up to 5,800 rpm. The crankshaft rpm is reduced 2.271:1 by a gear box integral with the front of the engine. Because of the high engine rpm, only automotive oil with its anti-foaming additives is approved for the Rotax 912. Another feature of this modern aircraft engine is the dual automatic altitude-compensating carburetors that provide automatic mixture control for ease of operation and continuous optimum fuel/air mixture for unsurpassed fuel economy. Unlike most other piston-powered airplanes, there is no pilot-adjusted mixture control in the Katana cockpit.

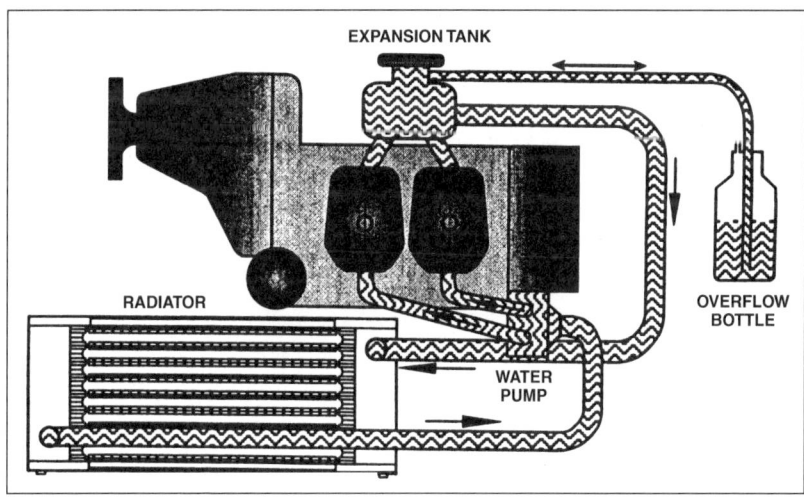

Cooling system overview

1-13

Diamond Katana DA20: A Pilot's Guide

Cooling system—radiator

The engine sets on hard rubber isolator mounts attached to a welded steel bar frame. The engine-mount frame is attached to the aircraft firewall using bolts and self-locking nuts. Attached to the engine-mount frame are different mounting brackets for oil and coolant radiators, oil tank, as well as mounting provisions for various cables and hoses for the engine. The engine-mount assembly is designed to reduce the transmission of vibration from the engine to the airframe.

DA20-C1

The powerplant for the DA20-C1 is the Teledyne Continental IO-240-B engine, with a 2,000-hour Time Between Overhauls (TBO). The IO-240-B is a downdraft, fuel injected, 4 cylinder, 4 stroke engine with horizontally opposed air-cooled cylinders and heads, a 239.8 cu. in. (3.9 liters) displacement. The propeller turns directly from the crankshaft. At a maximum rpm of 2800, the engine develops 125 h.p.

Section 1 **General Description**

At the recommended cruise rpm of 2550, the engine develops 95 h.p. The engine compression ratio is 8.5:1 with a bore of 4.44 inches and a stroke of 3.88 inches. The fixed-pitch propeller attaches to the engine by a doweled flange with six bolt holes. A propeller extension attaches between the propeller flange and the engine flange. The fuel control system is a Continental Continuous Flow Injector. The engine is mounted to the aircraft firewall with a tubular steel frame that incorporates rubber and metal bushings as shock mounts. The cowling is in two halves that attach to the airframe with camlock fasteners. The top cowling has a left and right air intake. The bottom cowling has one large air intake.

Continental IO-240-B engine in the DA20-C1

1-15

The Propeller
DA20-A1

The Rotax 912 engine/gear box output is transferred to a constant-speed two-bladed composite construction propeller. The propeller is a Hoffmann HO-V352F/170FQ that has a laminated hardwood core covered with high-strength carbon fiber composite material, incorporating an aluminum insert in the leading edge. The blade is protected from the weather by a high-quality polyurethane paint. The weight of the propeller and hub is only 12 pounds. The propeller rotates clockwise as seen from the cockpit, and has a diameter of 5 feet 6.9 inches (1.7 m).

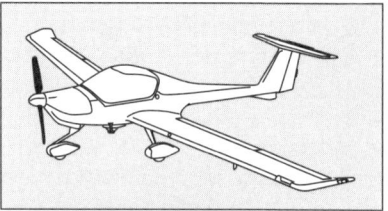

The propeller pitch is controlled by a Woodward hydraulic propeller governor, which is adjusted by a standard-type prop lever on the cockpit throttle quadrant, with a pitch angle range of 10-35 degrees. The propeller pitch directly determines the propeller rpm, within the limits of available engine power. Throttle determines engine power, and propeller pitch determines how efficiently that power is used. When a desired propeller rpm is selected, the governor automatically keeps the propeller speed at a constant value. The Katana is unique in having a constant-speed propeller to maximize the efficiency of a small engine for takeoff, climb and cruise. (A constant-speed propeller is usually found in higher performance airplanes.)

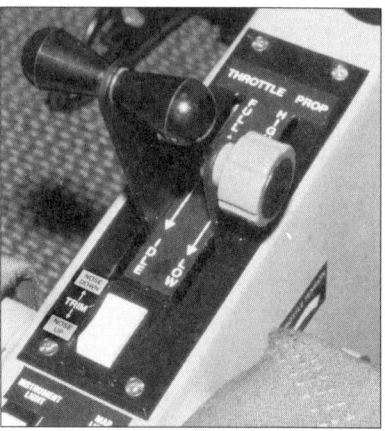

Throttle quadrant

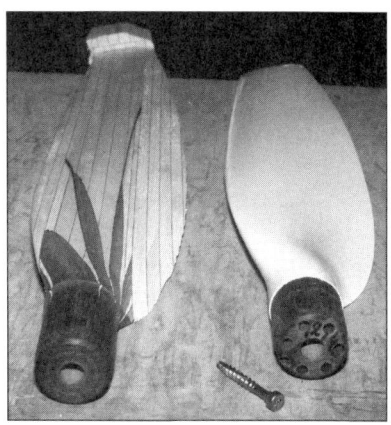

Section 1 **General Description**

The aluminum-alloy or Carbon Fiber Reinforced Plastic (CFRP) spinner assembly consists of the spinner dome, a front guide plate and a rear spinner bulkhead. The spinner is attached to the spinner bulkhead with ten screws.

Propeller governor

DA20-C1

The Continental IO-240-B powered DA20 was initially certified with a Hoffman (HO-14HN-175-157) composite-material, fixed-pitch propeller. This propeller has a blade pitch optimized for cruise flight. The C1 Type Certificate was amended to include a Sensenich (W69EK-63) fixed-pitch, composite-material, climb-optimized propeller as an approved option on this model Katana.

The type of propeller used on the airplane affects climb and cruise performance. Always consider the appropriate aircraft performance chart in the approved flight manual for the DA20-C1 Katana. Note that the C1 model does not use a constant-speed propeller because of the weight it would add to the front of the airplane which houses a heavier engine than the original A1 model.

Diamond Katana DA20: A Pilot's Guide

The Ignition System

DA20-A1

Unlike most aircraft engines with magnetos, the Bombardier Rotax 912 in the Katana utilizes a modern dual electronic ignition system. Specifically, this engine uses a capacitor discharge ignition (CDI) system with solid state electronics to provide spark to the spark plug, to ignite the fuel/air mixture in the engine cylinders. Power for the electronic ignition system is provided by an electrical charging system that is integral with a crankshaft flywheel, with its own generator stator for each ignition circuit. The ignition system operates with total independence from the rest of the aircraft's electrical system. Once the engine is running, it will operate regardless of the serviceability of the battery or the electrical system alternator. Note that electronic ignition systems are very efficient: there are no moving parts to wear and create errors in engine timing.

The Rotax 912 uses a capacitor discharge ignition (CDI) system.

DA20-C1

The Continental IO-240-B powered DA20 utilizes a conventional aircraft dual magneto ignition system. Two Unisom (slick) Model 4381 magnetos with impulse couplings supply the high voltage electrical pulse to the spark plugs. Each cylinder head has two spark plugs to give ignition to the engine.

The Continental IO-240-B powered DA20 utilizes a conventional aircraft dual magneto ignition system. Two Unisom (slick) Model 4381 magnetos with impulse couplings supply the high voltage electrical pulse to the spark plugs. A magneto uses a permanent magnet driven by the aircraft engine to generate electric current independent of the aircraft's electrical system. The impulse coupling on each magneto enables the unit to provide a surge of electrical current when the engine is started, and just as the crankshaft rpm is starting to spin up to its normal operating speed. Each cylinder head has

Section 1 **General Description**

two spark plugs to give ignition to the engine. The dual ignition system improves combustion efficiency and results in a slightly higher power output when the dual system is functioning. The two identical and independent systems provide the safety of redundancy if one of the magnetos fail.

The Oil System
DA20-A1

The engine's oil system provides lubrication, cooling, sealing, cleansing and protection against corrosion. The Bombardier Rotax 912 has a dry-sump, forced lubrication system with a separate oil tank located outside the engine. The oil tank, located on the right-hand side of the engine, has a capacity of 3.2 quarts. Oil flows from the crankcase to the oil tank, and then to the oil radiator. The oil tank is accessible through an upper cowling inspection hatch for easy checking of the oil level.

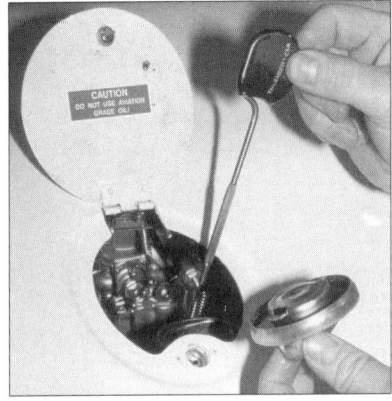

Checking the oil

Oil quantity should be checked after the engine has run at idle speed for one minute. The reason for this is that it is possible for some of the oil in the external oil tank to drain back into the engine if the engine has not been operated for a period of time. This will cause the oil level to indicate lower than actual conditions.

Oil pressure and oil temperature

An opening in the right lower cowling supplies ram air to the oil radiator for cooling. When the aircraft is operated in winter weather conditions with ambient temperatures below 32°F, a baffle must be installed at the oil cooler inlet. This allows the oil to warm up to normal operating limits. A fire sleeve to meet FAA fire resistance requirements covers all oil lines that connect the various components of the oil system.

An oil filter is located at the front of the crankcase adjacent to the oil pump. The oil temperature gauge mounted in the cockpit is electrically operated, and measures temperatures from a sending unit installed next to the oil filter. The oil pressure gauge in the cockpit gives an electrical reading from an oil pressure sensor installed opposite the oil filter.

Diamond Katana DA20: A Pilot's Guide

DA20-C1

The engine in the DA20-C1 is described as a conventional aircraft engine and as such uses only oil approved for aircraft engines. The engine uses a high-pressure wet pump lubrication system. The mechanical oil pump is located in the oil sump. The system includes an oil cooler. The oil level dipstick is marked for U.S. quarts and the operating level should be between 4 and 6 quarts.

The Starter System
DA20-A1

The Rotax 912 starter motor is located at the back of the lower right-hand side of the engine. The electric starter is a 12-volt, 0.6 kW DC motor. The drive pinion actuates via an intermediate free wheel gear on the crankshaft. By way of the free-wheel, torque is transferred to the crankshaft. After ignition and deactivation of the electric starter, the sprags of the sprag clutch return to their home position and the electric starter stops.

DA20-C1

The Continental IO-240-B engine uses an electrically operated 12-volt motor that attaches to the rear of the accessory drive case. During starting, the motor turns the engine through a reduction gear and a clutch. When you set the ignition switch to the start position, battery power energizes the starter motor. When the starter motor spins up, it pushes a pinion gear forward to engage the crankshaft gear. At this point the starter motor turns the pinion gear through a pair of reduction gears that turn the crankshaft. When the engine starts and the ignition switch is released, the pinion gear retracts from the crankshaft gear. A starter annunciator light comes on when the starter relay closes.

Section 1 **General Description**

The Fuel System
DA20-A1

The DA20-A1 Katana has a single fuselage-mounted aluminum fuel tank located behind the seats, below the baggage shelf. The fuel filler is on the left side behind the canopy. A fuel vent line runs from filler neck to the bottom of the fuselage, where the line is exposed to the exterior. The capacity of the tank was originally published as 20.3 U.S. gallons usable, but due to a manufacturing change the capacity is now listed as 19.5 U.S. gallons usable. The fuel lines run from the fuel tank to the electric fuel pump, then on to a fuel shut-off valve. The electric fuel pump incorporates a fuel filter (located under the fuel tank) and is accessible through an inspection panel on the belly of the airplane.

Fuel filler, left side behind canopy

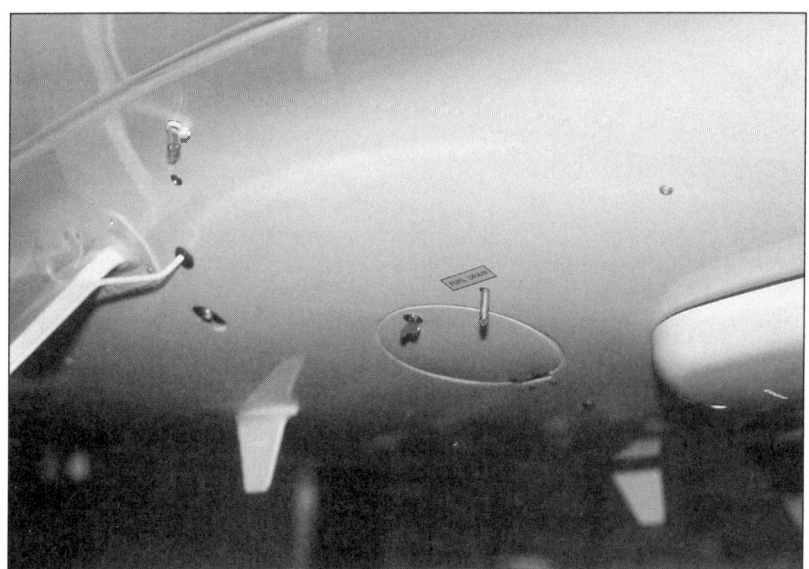

Fuel tank at bottom of fuselage, below baggage shelf—note fuel drain access tube

1-21

Diamond Katana DA20: A Pilot's Guide

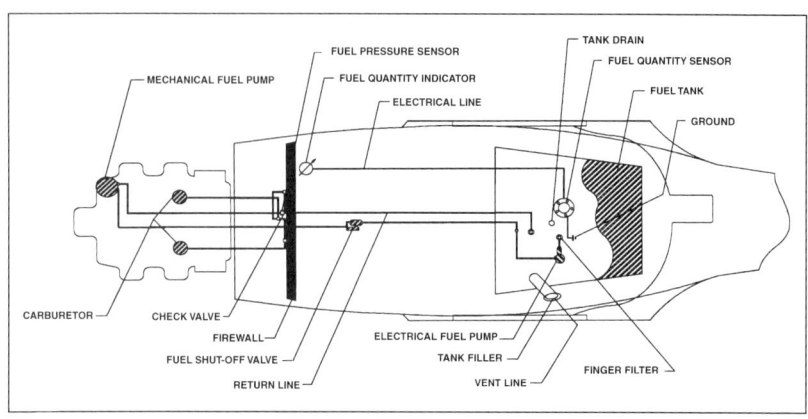

Fuel system diagram, DA20-A1

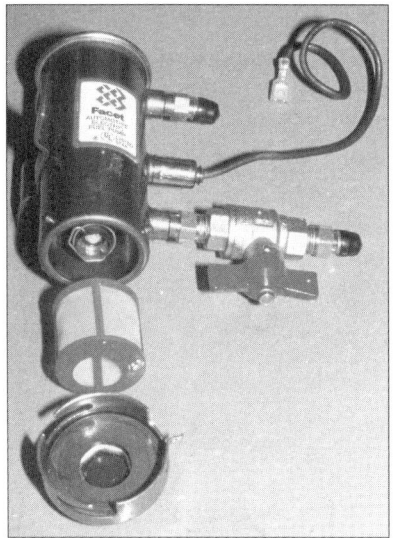

Fuel pump and filter

Section 1 **General Description**

The fuel shut-off valve is located in the cockpit on the left side of the center console and has an OPEN and CLOSED position. From the fuel shut-off valve the fuel line runs through the firewall breach fitting, and across the top of the engine to the mechanical fuel pump mounted on the right front side of the engine. From this engine-driven fuel pump, the fuel line continues back to a cross-shaped fitting on the firewall. It is on this fitting that the pressure sensor is attached, which activates the fuel pressure warning light if the fuel pressure drops below 1.45 psi.

From this cross-shaped fitting the fuel flow is directed to the two carburetors. Another line from this cross-shaped fitting returns fuel not required by the carburetors back through a check valve (which opens at 5 psi) to the fuel tank. The purpose of this check valve is to prevent vapor lock by maintaining proper fuel pressure. A fire-resistant fire sleeve covers all fuel lines for added safety in the engine compartment. (Note: to meet the FAA requirements as a fire resistant line, it must withstand exposure to a direct flame for five minutes under specified flow conditions without failure.)

A fuel quantity sensor is installed in the fuel tank and displays the fuel quantity on the panel-mounted fuel gauge. Since electric fuel gauges may malfunction, the quantity of fuel in the fuel tank can be visually checked using a graduated fuel pipette or wooden dowel supplied with the Katana. The graduated end of the pipette is dipped into the tank until it touches the bottom, then a finger is placed firmly over the top end of the pipette to hold the fuel in the tube as it is withdrawn for inspection. The fuel pipette is graduated in quarter tank increments. It is always good operating practice to visually check the fuel quantity prior to each flight. A fuel tank sump drain is provided to facilitate a preflight check and draining of fuel contaminants. This drain valve, located at the lowest point of the fuel tank, is activated by pushing up on a brass tube protruding from the left underside of the airplane. Fuel from this drain should be collected in a fuel sampler cup and visually checked. After draining fuel, make sure the valve has re-seated and does not drip or continue to drain fuel.

DA20-C1

The fuselage portion of the fuel system for the DA20-C1 is very similar to the DA20-A1 except the fuel capacity has been increased to 25 U.S. gallons (95 liters or 20.9 imperial gallons), with 21.3 gal (80.5 liters or 17.7 imp. gal.) usuable. In this system, 3.8 gal (14.5 liters or 3.2 imp. gal.) are unusable. The amount of unusable fuel can be reduced to 1/2 gal (2 liters or 0.4 imp. gal.) and the total usable increased to 24 gal (91 liters or 20 imp. gal.) if Service Bulletin DAC1-28-01 has been incorporated in the aircraft.

The fuel shut-off valve is located on the right side of the center console of the cockpit. Note that in the C1, the electric fuel pump is for priming the fuel injectors only—it is not an emergency standby fuel pump. Turning the pump on while the engine is running will enrich the mixture considerably. Although the effect is less noticeable at high power settings when the fuel flow rate is high, the effect at low and idle throttle settings will be an overrich mixture, which may cause rough engine operation. Therefore, the C1 approved flight manual recommends that for normal operation, the priming pump be turned off.

The Carburetors
DA20-A1

The purpose of the carburetor is to mix air with the fuel from the fuel system and supply the fuel/air mixture to the cylinders. The Rotax 912 uses dual Bing 64/32 constant-velocity, altitude-compensating carburetors (see the photograph on Page 1-14). Air for the dual carburetors enters through an intake scoop on the lower left cowling. This air is filtered and then fed into a balance tube from which the two carburetors draw air. A domed chamber at the top of the carburetor has a diaphragm that responds to atmospheric pressure change by moving a fuel needle. For the pilot this means that unlike most piston airplanes, the Diamond Katana does not have a cockpit mixture control. This reduces the pilot workload because there's no mixture to manually adjust with changes in altitude. The altitude-compensating carburetors also result in better fuel economy and less spark plug fouling, because pilots are prevented from flying on the overly-rich side of the fuel/air mixture. Problems with over-leaning are prevented because the engine runs on the mixture setting provided by the manufacturer.

From the carburetor, the fuel/air mixture is carried to the induction manifold, which feeds the fuel/air mixture to the intake port of each cylinder. A manual choke is located on the cockpit forward center console to enable the pilot to enrich the fuel mixture when cold-starting the engine.

Section 1 **General Description**

Technically the Rotax 912 uses a constant velocity, side-draft carburetor. When air flows through the carburetor, its velocity increases and both its pressure and temperature decrease. When the liquid fuel is discharged into the throat of the carburetor, it vaporizes and this change also decreases the temperature. The functions going on within the carburetor are processes that result in cooling the throat of the carburetor. All carburetors have some degree of proneness to developing, under certain power, temperature, and humidity conditions, ice that could block the throat of the carburetor. Because of the possibility of carburetor ice choking off air to the engine, a provision is made for a pre-heated air supply to the carburetor. In the cockpit a "pull on – push off" carburetor heat knob is located on the lower center of the instrument panel forward of the throttle quadrant (*see* photograph on Page 1-32). The use of this control and the subject of carburetor icing are discussed in Section 4.

The fuel system carburetors on the Rotex 912 engine do not use a manual fuel primer, but instead use a manual choke for cold-engine starts. Pulling on the manual choke enriches the fuel/air ratio and increases the idle rpm. For cold-engine starts the throttle must be pulled all the way back to the idle position when the choke is pulled, or the fuel/air mixture will be too rich for the engine to start properly. The choke is spring-loaded to return back to the "OFF" position and must be physically be held "ON." The starting choke is not used if the engine is warm. When starting the engine it is important that the throttle friction is sufficient to hold the throttle in a desired position. The throttle friction can be adjusted with the tension knob on the right side of the center console.

DA20-C1

The Continental IO-240-B is a fuel-injected engine. Fuel is metered to the engine by a device called a throttle body. It then travels to the fuel distribution manifold, mounted on top of the engine. The distribution manifold divides the fuel flow to each of the four injector nozzles at the intake port. A return line and a check valve runs from a fuel vapor separator to the fuel tank. A fuel pressure sensor is installed at the fuel distribution manifold. Since the engine does not use a carburetor, there is no need for carburetor heat to deal with carburetor ice.

DA20-C1 fuel injection

1-25

The Electrical System

The DA20 Katana is equipped with a 14-volt direct current electrical system. A 40-amp alternator is mounted on the forward left side of the engine and is belt-driven from the propeller hub. A 12-volt, 20-amp-hour lead acid battery is located on the forward left side of 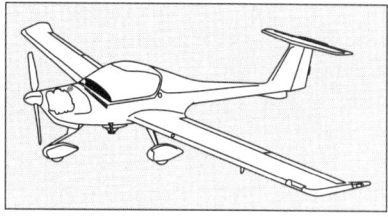 the firewall in the DA20-A1 and aft of the baggage compartment in the DA20-C1. The pilot controls the electrical system via the MASTER SWITCH, which is red in color, located on the lower left side of the instrument panel. An overvoltage sensor protects the electrical system in the event of system overvoltage.

With the exception of the battery, alternator, and starter motor, all electrical components within the engine compartment are mounted to the electrical shelf, which is attached to the engine mount. Connections to the cockpit are established through two terminal strips on the electrical plate and then hardwired through the firewall. The status of the electrical system is presented to the pilot by the generator/alternator warning light, ammeter, voltmeter, and automatic circuit breakers located on the right-hand side of the instrument panel.

Belt-driven alternator

Section 1 **General Description**

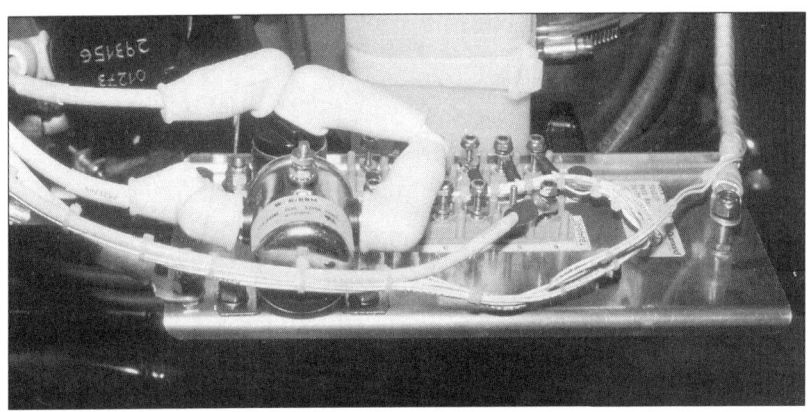

Engine electrical components mount to electrical shelf

The ALTERNATOR is the primary source of power for the aircraft's electrical consumers in normal operation with the engine running. The alternator produces alternating current (AC), which is converted into direct current (DC) by diodes incorporated into the alternator housing. These diodes act as rectifiers and serve as the voltage regulator. The voltage regulator is adjusted by the manufacturer to a fixed threshold of 14.0 volts; this cannot be readjusted. An overvoltage sensor provides protection in the case that, due to a malfunction, voltage increases above 16.1 volts. By design, alternators require a small voltage (about 3 volts) to produce the electromagnetic field required inside the alternator. The significance of this is, if the battery is completely discharged (flat), the alternator will not be able to supply power to the electrical system, even after the engine has been started by some other means (i.e., external power or hand-propping).

The primary purposes of the BATTERY are to provide power for the engine starting, the internal excitation of the alternator, and as a backup in the event of alternator failure. In normal operations with the engine running, the alternator provides the power to the electrical consumers and charges the battery. A fully charged battery has a charging current of about 2 amperes; in a partially discharged condition (i.e., just after engine start) the charging rate can be much higher than this. In the event of an alternator failure, the battery is providing all power to the electrical consumers.

In theory, a fully charged 20-ampere-hours battery is capable of providing 20 amps for 1 hour, 1 amp for 20 hours, or 10 amps for 2 hours, etc. In practice, the power available is governed by factors such as battery age and condition, load placed on it, and so on. In the case of a charging system failure, the best recourse is to reduce electrical load to the minimum consistent with safety, and plan to land at the earliest opportunity.

Diamond Katana DA20: A Pilot's Guide

The AMMETER is located on the right-hand section of the instrument panel; it indicates charging (+) and discharging (-) of the battery. A positive deflection usually indicates that the battery is being charged. A negative needle deflection indicates that the battery is being drawn upon to power the electrical consumers. Normally the needle should rest near the center or zero mark. It should be noted that after starting the engine, the battery has been drained slightly and the ammeter should show a positive charge as the alternator is replenishing the battery. Should the alternator charging system fail, the ammeter will show a discharging needle to the left indicating that electrical current is being drawn out of the battery to power the electrical consumers.

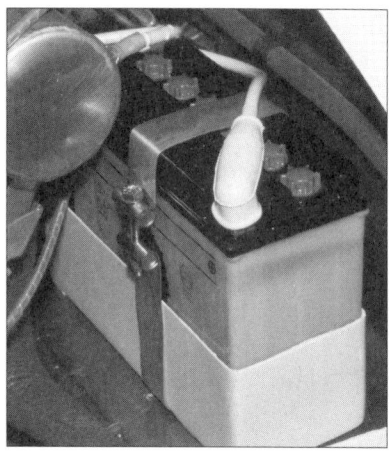
Battery installation, DA20-A1

The VOLTMETER is located above the ammeter on the right-hand section of the instrument panel. This instrument gives direct information about the alternator output or the amount of energy available in the battery. The display on the face of the voltmeter is subdivided into red, yellow and green sections. If the meter is indicating a reading in the yellow region, this indicates that current supplied by the alternator is not sufficient for the demand. The voltmeter is the best indicator of the status of the electrical system.

An ALTERNATOR WARNING LIGHT (on some early serial number aircraft this light is labeled as generator warning light) is located in the top center section of the instrument panel. This red warning light is illuminated when the alternator is not producing power.

Voltmeter and ammeter

Section 1 **General Description**

The pilot controls the electrical system via the MASTER SWITCH. This switch is a split rocker switch with two halves, labeled BAT and ALT. Normally the switch is operated as one, with both halves used together. The BAT half of the switch can be operated independently, so that all electrical power is being drawn from the battery only; however, the ALT side can only be turned on in conjunction with the BAT half. Should a problem occur with the voltage regulator (overvoltage sensor), the condition can be interrupted by turning the master switch off for 2 seconds and then turning it on again to reset the system.

The DA20 Katana has a split electrical bus bar to distribute power to the electrical accessories and the avionics via automatic circuit breakers. Switches for electrical components such as the electric fuel pump and external lights are located on the bottom left side of the instrument panel. A separate AVIONICS POWER SWITCH is installed to control the avionics bus bar separately from the rest of the electrical consumers. This switch is normally OFF before engine start, in order to prevent the avionics from being damaged by transient high voltage during the engine start.

Right-hand instrument panel: circuit breakers

The various electrically operated systems are protected by individual CIRCUIT BREAKERS on the right-hand section of the instrument panel. Should a problem occur (i.e., a short circuit) the relevant circuit breaker may "pop," and will be raised in relation to the other circuit breakers. The correct procedure is to allow the circuit breaker to cool for 2 minutes, then reset by pushing the circuit breaker back in and check the results. If the circuit breaker pops again, it should not be reset. All circuit breakers show their ampere rating and the components they protect.

1-29

Diamond Katana DA20: A Pilot's Guide

Left-hand instrument panel: electric fuel pump, lights, avionics power, and master switch

Apart from engine starting and the alternator/battery field, the electrical system supplies power to the following:

- All internal and external lights
- All radios and intercom
- Wing flaps and elevator pitch trim
- Gyroscopic instruments
- Fuel, oil pressure, oil temperature gauges
- Electric fuel pump

The Katana can be fitted with an external power receptacle, by special order only.

The Stall Warning System

An intake with a red circle around it, in the left wing leading edge, sends an aural warning through a horn and reed located in the left corner of the instrument panel. When the stalling angle of attack is approached, the airflow over the leading edge causes suction through the reed producing a loud tone, which becomes increasingly high-pitched as the stalling angle of attack is reached. Typically, the stall warning activates 5-10 knots above the stall speed.

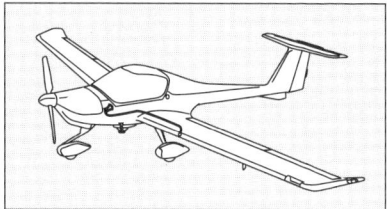

Leading edge stall warning intake

Section 1 **General Description**

To check the function of the warning horn on the ground, the intake should be checked for blockages by sucking air through the intake. It is recommended that a handkerchief or something similar be placed over the intake first, to avoid the unpleasant possibility of swallowing insects through the intake.

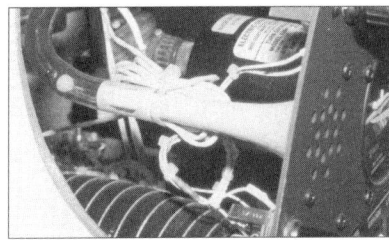

Cockpit stall horn behind panel

The Lighting System

The exterior lights of the DA20 Katana include anti-collision lights (ACL), position lights, landing and taxiing lights. The anti-collision strobe light and position lights are combined into one unit located at each wing tip. The power supply for the strobe lights is installed under the pilot's seat. The landing and taxiing lights are installed in the leading edge of the left wing near the wing tip. Exterior lights can be individually selected from a row of rocker switches on the lower left side of the instrument panel (*see* the photograph on Page 1-30).

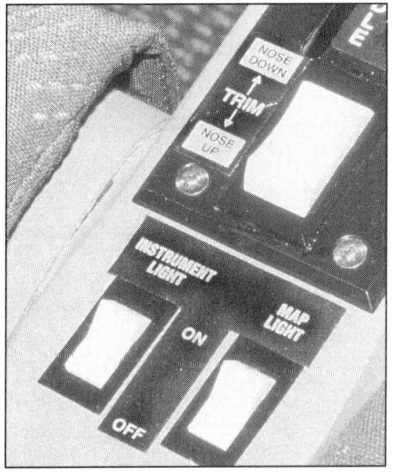

Switches for panel "flood" and map lights

Cockpit lighting of the DA20 Katana is provided by a flood-lighting module located at the top of the cockpit, above and behind the pilot's head and on the centerline of the aircraft. Incorporated into the cockpit flood-lighting module are two instrument panel illumination lights and one map light. The switches for the panel and map lights are located on the center console aft of the trim control switch. A dimming rheostat for adjusting the intensity of the instrument flood-lighting is located in the upper left corner of the instrument panel. There is also a toggle switch near the top center of the instrument panel that controls the intensity of the wing flap and trim position indicators. A red light mounted under the left side of the instrument panel illuminates the Fuel Shutoff Valve.

Diamond Katana DA20: A Pilot's Guide

Instrument panel

The Avionics

The radio and navigation equipment is located in the center portion of the instrument panel. The flat panel of the Katana provides space for a complete IFR training package that may include Audio Panel with Marker Beacons, dual Nav/Com radios, ADF, DME, Transponder, and GPS navigation system. A voice-activated intercom is standard equipment. A push-to-talk for radio transmissions is installed in the control stick. There are two sets of connectors for headsets located on the backrest of the seats.

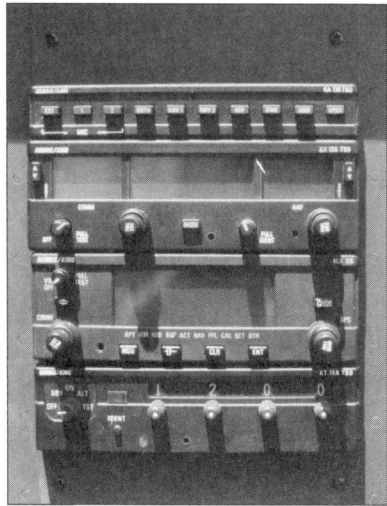

Section 1 **General Description**

The Gyroscopic Instruments

Three of the six flight instruments are gyroscopic instruments: the attitude indicator or artificial horizon, the heading indicator, and the turn indicator. Typically in light general aviation airplanes the attitude indicator and the heading indicator are vacuum-powered gyros. However, unlike most light aircraft engines, the Rotax 912 installation in the DA20-A1 does not provide an accessory outlet for a vacuum pump. In the A1 model all gyroscopic instruments are electric, using individual electric motors to spin the gyros at high speed, rather than an engine-driven vacuum system. The Continental-powered DA20-C1 has an engine-driven vacuum pump to power the attitude indicator and the heading indicator. In the C1 model a suction gage is located in the upper left corner of the instrument panel.

Detail of left side, instrument panel, DA20-A1

1-33

Diamond Katana DA20: A Pilot's Guide

The Pitot-Static System

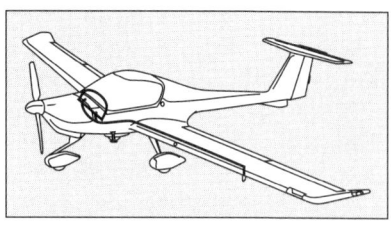

The pitot-static system provides static pressure to the vertical speed indicator (VSI) and altimeter, and both dynamic pressure and static pressure to the airspeed indicator. The Katana utilizes a combination pitot tube and static port located under the left wing. Impact pressure from ram air is measured from the front side of the tube, while static pressure is sensed through the static port located on the backside of the tube. This type of combination pitot-static also provides a small drain hole at the bottom to allow water that may have entered the pitot tube to escape. For additional protection against water and humidity, water sumps are installed within the pitot static lines. These water sumps are accessible beneath the left seat pan.

The pitot tube should be checked before flight to ensure that the pitot and static ports are unobstructed. The pitot tube may be protected on the ground with a removable pitot cover. It is important not to blow into either pitot or static vents as this can result in damage to the pressure instruments.

The Katana pitot static tube does not incorporate a heating element as is common in similar pitot-static tubes in other aircraft. It must be remembered that the DA20 Katana is not approved for flight into known icing conditions. No checking system is incorporated into the aircraft's system; moreover, instrument indications in the event of a leak or blockage are

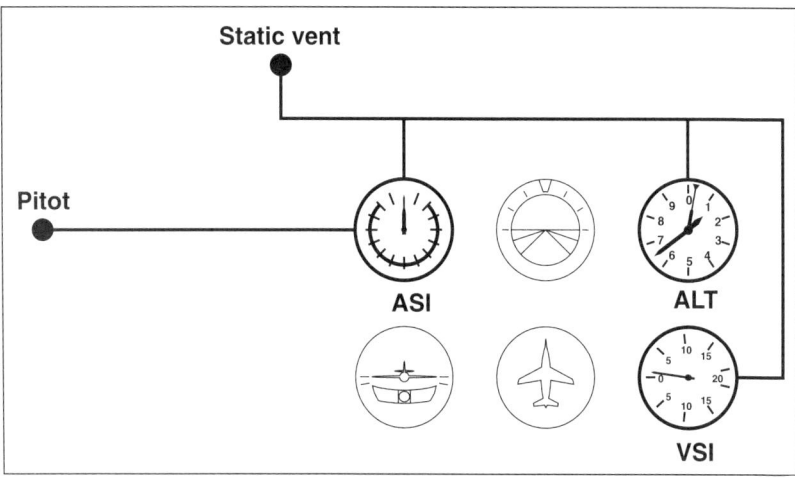

Section 1 **General Description**

outside the scope of this book. Should a static line become blocked, it is possible to provide static pressure for the system by breaking the face of the vertical speed indicator, allowing pressure from the cabin to enter the system. This action is rather drastic, probably requiring the use of the fire extinguisher, and should be considered a last resort.

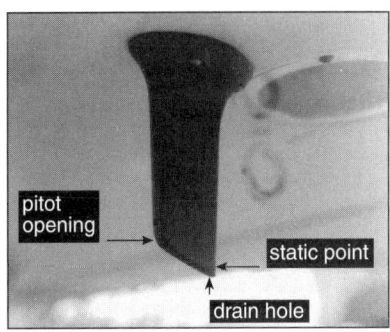

Pitot-static mast

The Heating and Ventilation System

The cabin heating and canopy defrost is the result of ram air directed through the coolant radiator and into a heat shroud located around the muffler, and into a pilot-controllable valve. The flow of heated air can be regulated with a cabin heat, pull-on/push-off knob located on the lower center of the instrument panel, just forward of throttle quadrant. Two fuselage intake vents that bring fresh air to adjustable nozzles are located on each side of the instrument panel, and provide cockpit ventilation. An exhaust vent in the baggage compartment provides ventilation flow-through for enhanced air movement. The two sliding windows in the canopy can be opened for additional ventilation.

Cockpit exhaust—reverse NACA air vent

Cabin heat knob on panel forward of throttle quadrant, DA20-A1

1-35

Diamond Katana DA20: A Pilot's Guide

Seats, Harnesses and Baggage Compartment

The Katana seats are an integral part of the cockpit structure and provide a very comfortable semi-reclined seating position. People of different heights are accommodated by adjusting the rudder pedal distance fore or aft. The pedals for the rudder and brakes are adjusted by pulling the "T" handle located at the front of the seats or at the front of the rudder pedal assembly, to unlock and move the pedal distance. For safety reasons, the pedals should only be adjusted on the ground. Seat cushions rest on the molded seat pans whose sides form an armrest. The seat pans themselves serve as removable inspection panels to facilitate maintenance and inspection of the underlying components. Each seat is provided with a four-point safety belt that can be individually adjusted.

Seat harness

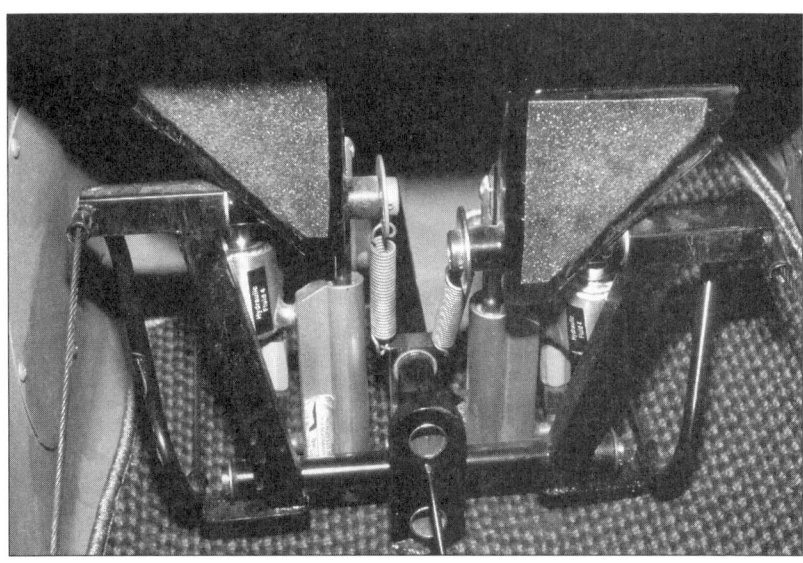

Rudder pedal adjustment mechanism

Section 1 **General Description**

The baggage compartment for the Katana is located behind the seat above the fuel tank. Maximum weight in the baggage compartment is placarded to be 44 lbs/20 kg. The baggage net that separates the cargo area from the cockpit should always be secured prior to taxi and flight. A fire extinguisher, rescue hammer (on C1 model), emergency locator transmitter and aircraft document pouch are located in the baggage compartment.

The Canopy

One of the attractive features of the Katana is the unsurpassed cockpit visibility on the ground and in the air. The single-piece acrylic canopy is mounted in a glass and carbon-fiber reinforced plastic frame. The canopy lifts up on well-balanced swivel arms and provides ease of entry to the cockpit. Locking the canopy once you are inside the airplane is simply a matter of pulling the canopy closed and pushing forward the red latches on the left and right side of the canopy frame. A canopy warning light is located near the top center of the instrument panel and is illuminated if the canopy is not locked properly. To open the canopy, both left and right latches must be slid back independently; however, an emergency release lever on the left side will open the right latch. Left and right sliding side windows are installed as a standard feature. A canopy lock on the left side is also a standard feature.

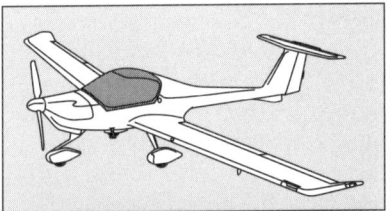

Katana's canopy lifts up on swivel arms providing easy cockpit entry.

Canopy latch warning light

1-37

Diamond Katana DA20: A Pilot's Guide

Section 2
Limitations

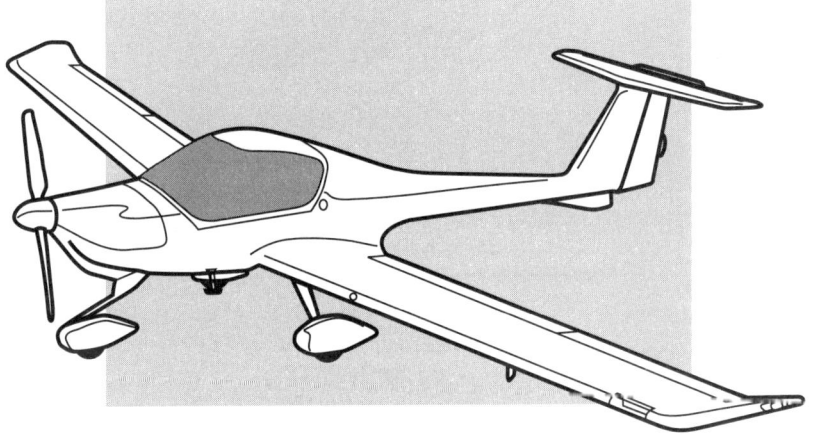

Diamond Katana DA20: A Pilot's Guide

Section 2 **Limitations**

Diamond Katana DA20-A1 Dimensions

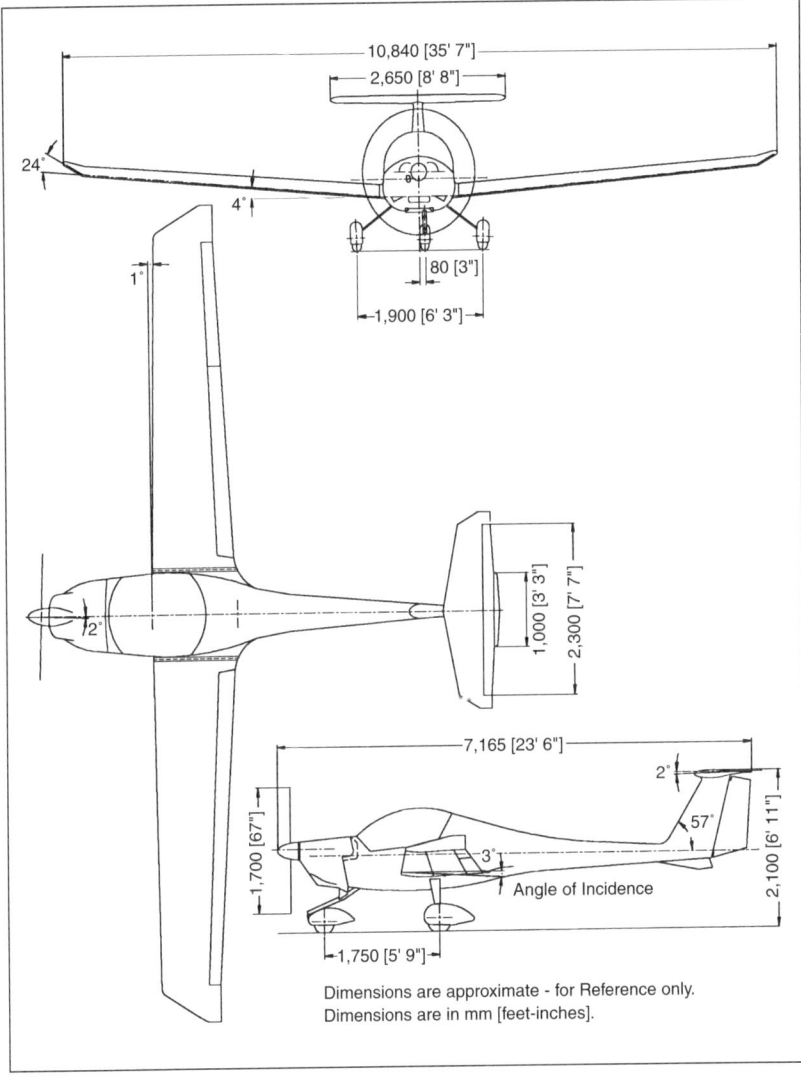

Dimensions are approximate - for Reference only.
Dimensions are in mm [feet-inches].

Diamond Katana DA20: A Pilot's Guide

Diamond Katana DA20-C1 Dimensions

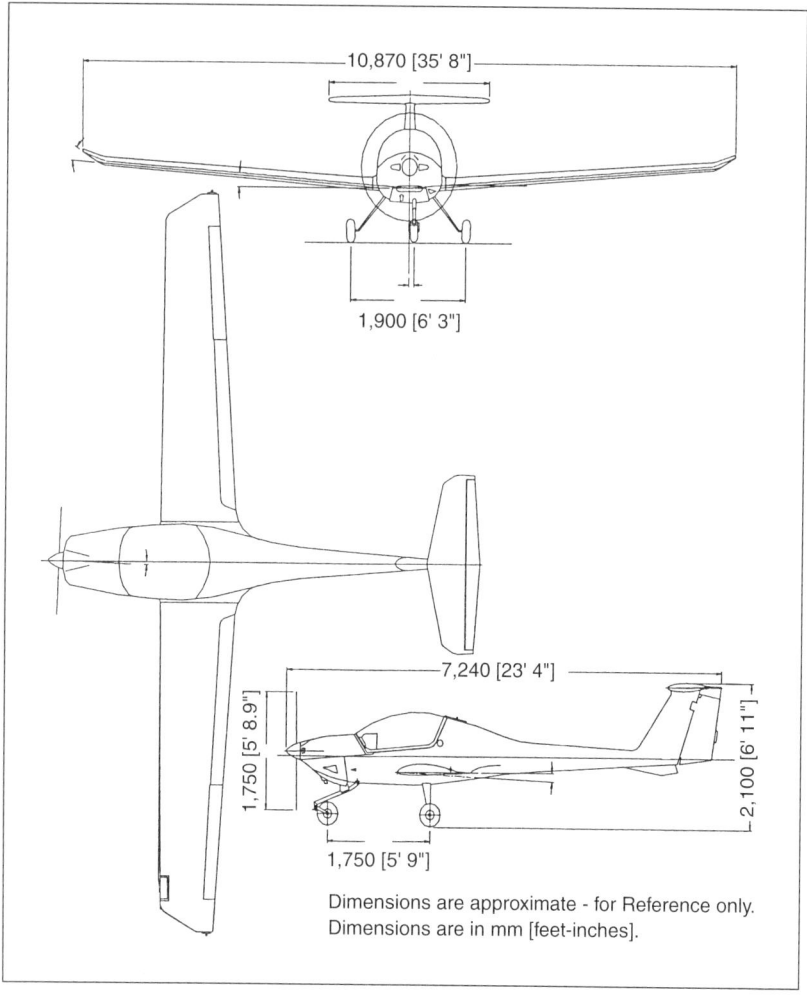

Section 2 **Limitations**

The "V" Airspeed Code

V_{S0} — (Low end of white arc) Stalling speed with full flaps.

V_{S1} — (Low end of green arc) Stalling speed without flaps.

V_{FE} — Maximum airspeed with flaps extended. Do not extend flaps above this speed, or fly faster than this speed with any flaps extended.

V_A — Design maneuvering speed. Do not make full or abrupt control movements when flying faster than this speed. Design maneuvering speed should not be exceeded when flying in turbulent conditions.

V_{NO} — Maximum structural cruising speed. Do not exceed this speed except in smooth air conditions.

V_{NE} — Never exceed speed. Do not exceed this airspeed under any circumstances.

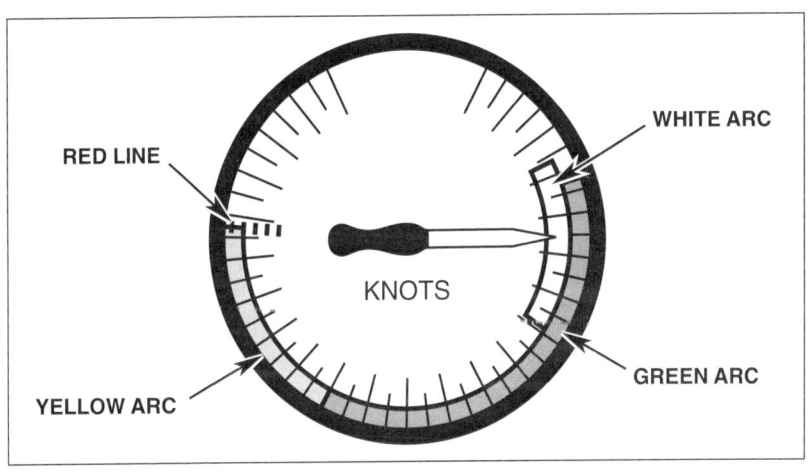

2-5

Diamond Katana DA20: A Pilot's Guide

DA20-A1

Airspeed Limitations (All quoted speeds are INDICATED airspeed—IAS)

	Knots	mph	km/h
V_{NE}	161	189	304
V_{NO}	118	135	218
V_A	104	122	196
V_{FE}	81	90	144
Stalling Speed clean	41	48	78
Stalling Speed Full Flaps	37	39	63

Airspeed Indicator Markings

	Knots	mph	km/h
Red Line (Never Exceed)	161	189	304
Yellow Arc (Caution range)	118 – 161	135 – 189	218 – 304
Green Arc (Normal operating range)	41 – 118	47 – 135	76 – 218
White Arc (Flaps extended range)	37 – 81	39 – 90	63 – 144

Maximum Demonstrated Crosswind Component   15 knots   17 mph   27 km/h

Airframe Limitations
Weights

Maximum Permissible Weight	1,609 lbs	750 kg
Maximum Baggage Weight	44 lbs	20 kg

Temperature Limits

Minimum Takeoff Temperature (OAT):	-13°F	-35°C
Maximum Takeoff Temperature (OAT):	131°F	55°C

DA20-C1

Airspeed Limitations (All quoted speeds are INDICATED airspeed—IAS)

	Knots	mph	km/h
V_{NE}	164	189	304
V_{NO}	118	135	218
V_A	106	122	196
V_{FE}	78	90	144
Stalling Speed clean	42	48	78
Stalling Speed Full Flaps	34	39	63

Airspeed Indicator Markings

	Knots	mph	km/h
Red Line (Never Exceed)	164	189	304
Yellow Arc (Caution range)	118 – 164	135 – 189	218 – 304
Green Arc (Normal operating range)	42 – 118	47 – 135	76 – 218
White Arc (Flaps extended range)	34 – 78	39 – 90	63 – 144

Maximum Demonstrated Crosswind Component   15 knots   17 mph   27 km/h

Airframe Limitations

Weights

Maximum Permissible Weight	1,653 lbs	750 kg
Maximum Baggage Weight	44 lbs	20 kg

Temperature Limits

Minimum Takeoff Temperature (OAT):	-31°F	-35°C
Maximum Takeoff Temperature (OAT):	131°F	55°C

Diamond Katana DA20: A Pilot's Guide

DA20-A1 and DA20-C1
Flight Load Factors

	Flaps up	Flaps down
Positive	4.4	2.0
Negative	2.2	0.0

Performance Limitations

The published maximum cruising altitude is 13,120 ft (4,000 meters); however, this altitude is not published as a limitation in the *Airplane Flight Manual*.

Approved Maneuvers

Aircraft is certified for utility category maneuvers.

All normal flight maneuvers

Stalls (except dynamic stalls)

Lazy Eights, entry speed:	116 Kts	133 mph	215 km/h
Chandelles, entry speed:	116 Kts	133 mph	215 km/h

Steep turns, bank angle not to exceed 60 degrees

Spins (flaps up)

Kinds of Operation

Day/Night VFR

(Note: Aircraft S/N 10002 through 10020 are limited to day VFR operation only, unless Service Bulletin 95-01 has been complied with.)

Minimum Equipment List

(Note: There may be additional operational or national requirements.)

Flight and Navigation Instruments

Airspeed Indicator

Altimeter

Magnetic Compass

Turn and Bank Indicator (not mandatory for Day-VFR only)

Instrument Panel and Map Lighting (not mandatory for Day-VFR only)

Powerplant Instruments

Fuel Quantity Indicator

Oil Pressure Indicator

Oil Temperature Indicator
Manifold Pressure Indicator
Cylinder Head Temperature Indicator
Tachometer
Fuel Pressure Warning Light
Voltmeter
Ammeter
Generator Warning Light

DA20-A1
Engine Limitations (Rotax 912)

	Tachometer	Instrument Marking
Maximum rpm	2550	Red line
Normal Operating Range	950 – 2420	Green Arc
Caution Range	2420 – 2550	Yellow Arc
	Oil Temperature Indicator	**Instrument Marking**
Oil Temp:		
Lower Limit	122°F/50°C	Red Line
Normal Operating Range	122 – 284°F/ 50 – 140°C	Green Arc
Upper limit	284°F/140°C	Red Line
	CHT	**Instrument Marking**
Cylinder Head Temp	302°F/150°C	Red Line
	Oil Pressure Indicator	**Instrument Marking**
Oil Pressure:		
Lower Limit	22 psi/1.5 bar	Red Line
Normal Limit	22 – 73 psi/1.5 – 5 bar	Green Arc
Caution Range	73 – 102 psi/5 – 7 bar	Green Arc
Caution Range	73 – 102 psi/5 – 7 bar	Yellow Arc
	Voltmeter	**Instrument Marking**
Voltage:		
Lower Limit	8 – 11 volts	Red Line
Caution Range	11 – 12.5 volts	Yellow Arc
Normal Range	12.5 – 16 volts	Green Arc
Upper Limit	16.1 volts	Red Line

Diamond Katana DA20: A Pilot's Guide

Oil Capacity

	US Quarts	Liters
Minimum	2.1	2.0
Maximum	3.2	3.0

Oil Grade

Use only name-brand automotive mineral oil or semi-synthetic marked "SF" or "SG." When operating with AVGAS 100LL fuel, do not use synthetic oil. Oil is available in different grades, used according to the surface air temperature range in which the aircraft will be operating, and identified by SAE numbers. The table below shows the recommended grades for various temperature ranges.

Surface Air Temperature	Single Grade Oil	Multi Grade Oil
40 – 100°F/4.4 – 37.7°C	SAE 40	SAE 20W-40 or 50
05 – 65°F/-15 – 18.3°C	SAE 20	SAE 15W-40
05 – 100°F/-15 – 37.7°C		SAE 15W-40 or 50
-10 – 85°F/-23 – 29.4°C		SAE 10W-30

Coolant Type

Standard name-brand automotive antifreeze concentrate with additives against corrosion. The coolant should be diluted in a 80 to 20 ratio with distilled water for optimal temperature protection.

Coolant Capacity	US Quarts	Liters
Minimum	2.5	2.4
Maximum	2.6	2.5

Fuel Grade

The Diamond Katana DA20-A1 is certified for use with AVGAS 100LL or unleaded automotive gasoline with a minimum 90 RON or 87 AKI. Use of automotive gasoline containing alcohol (methanol or ethanol) other than anti-ice additives, is not permitted unless Service Bulletins DA20-7303 and DA20-7304 have been complied with.

Fuel Capacity	US Gallons	Liters
Total Quantity	20.1	79
Usable Fuel	19.5	74
Unusable Fuel	00.6	.2

Section 2 **Limitations**

Miscellaneous Limitations

Nose Wheel Tire Pressure

    26 psi/1.8 bar   4.00-4 S/N 10001 – 10050
                               5.00-4 S/N 10051 and subsequent

Main Wheel Tire Pressure

    33 psi/2.3 bar 15 X 6.00-5

DA20-C1
Engine Limitations (Continental IO-240-B)

	Tachometer	Instrument Marking
Maximum rpm	2801	Red Line
Normal Operating Range	700 – 2800	Green Arc
Oil Temp:	**Oil Temperature Indicator**	**Instrument Marking**
Lower Limit	75°	Red Line
Caution Range	75° – 170° & 220° – 240°	Yellow Arc
Normal Operating Range	170° – 220°	Green Arc
Upper Limit	240°	Red Line
Cylinder Head Temp:	**CHT**	**Instrument Marking**
Lower Caution Limit	240° – 360°	Yellow Arc
Normal Operating Range	360° – 420°	Green Arc
Upper Caution Limit	420° – 460°	Yellow Arc
Upper Limit	460°	Red Line
Oil Pressure:	**Oil Pressure Indicator**	**Instrument Marking**
Lower Caution Limit	10 – 30 psi	Yellow Arc
Normal Operating Range	30 – 60 psi	Green Arc
Upper Caution Limit	60 – 100 psi	Yellow Arc
Upper Limit	100 psi	Red Line
Fuel Pressure:	**Fuel Pressure Indicator**	**Instrument Marking**
Lower Limit	3.5 psi	Red Line
Upper Limit	16.5 psi	Red Line

Diamond Katana DA20: A Pilot's Guide

Voltage:	Voltmeter	Instrument Markings
Lower Limit	8 – 11 volts	Red Arc
Normal Operating Range	12.5 – 16 volts	Green Arc
Caution Range	11 – 12.5 volts	Yellow Arc
Upper Limit	16.1 volts	Red Line

Oil Capacity	**U.S. Quarts**	**Liters**
Minimum	4.0	3.78
Maximum	6.0	5.68

Oil Grade

Use only lubricating oils conforming to TCM specifications MH524. Almost all major brand aviation single-grade or multi-grade oils are applicable. The table below shows the recommended grades for various temperature ranges.

Surface Air Temperature	Single Grade Oil	Multi Grade Oil
40 – 100°F/4.4 – 37.7°C	SAE 40	SAE 20W-40
05 – 65°F/-15 – 18.3°C	SAE 20	SAE 15W-40
05 – 100°F/-15 – 37.7°C		SAE 15W-40 or 50
-10 – 85°F/-23 – 29.9°C		SAE 10-30

Fuel Grade

The Katana DA20-C1 is certified to use only AVGAS 100 or 100LL.

Fuel Capacity	U.S. Gallons	Liters
Total Quantity	25.0	95
Usable Fuel	21.3	80.5
Unusable Fuel	3.8	14.5

Miscellaneous Limitations

Nose Wheel Tire Pressure
   26 psi/1.8bar   500 – 4.6 ply

Main Wheel Tire Pressure
   33psi/2.3bar   500 – 5.6 ply

Section 3
Handling the Katana DA20

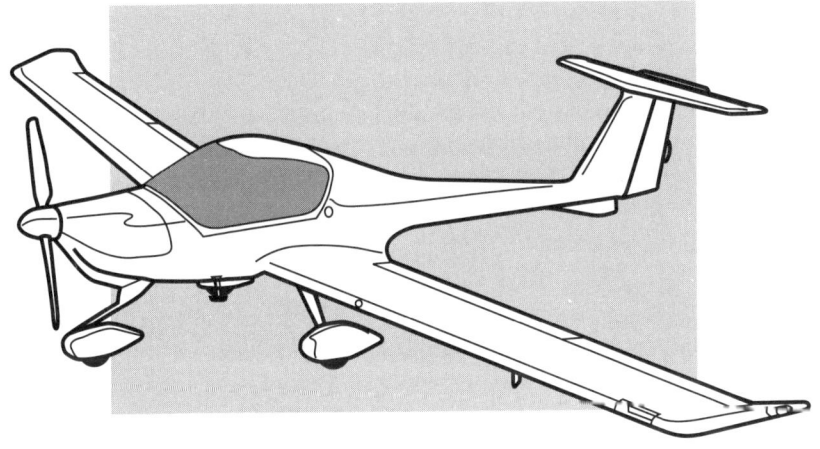

Diamond Katana DA20: A Pilot's Guide

Section 3 **Handling the Katana DA20**

Ground Handling

Note: *The information in this section is not a substitute for flight instruction under the guidance of a flight instructor familiar with the aircraft and its characteristics.*

If you follow simple instructions, the Katana is very easy to maneuver by hand in the parking ramp area or hanger. Diamond Aircraft provides a towbar, which attaches to the nosewheel assembly and provides a good point at which to push or pull the aircraft. The early DA20 Katanas had a nose wheel turn limit of 30 degrees; all *current* production aircraft have a turn limit of 64 degrees. Pulling the Katana forward can be assisted by pulling on the propeller at the root, next to the spinner. According to the Katana's *Airplane Flight Manual* (AFM), pushing or pulling the airplane by hand using the root of the propeller (with or without the use of the towbar) is approved.

Always remember safety considerations when handling a propeller. If a towbar is not used, the nose wheel will follow the movement of the airplane—which can create a challenge, particularly when moving the Katana backwards. The wing tips may also be used for pushing or pulling the DA20 Katana, according to the AFM. The real secret to moving the Katana backwards (and it is really no secret as it's published in the AFM) is to simply push down on the aft section of the fuselage near the vertical stabilizer, and raise the nose wheel completely off the ground. With the nose wheel off the ground, push back on the leading edge of the horizontal stabilizer close to the center. Using this technique, you can easily push the Katana DA20 from side to side as well as backwards. Do not push on the spinner or any control surface.

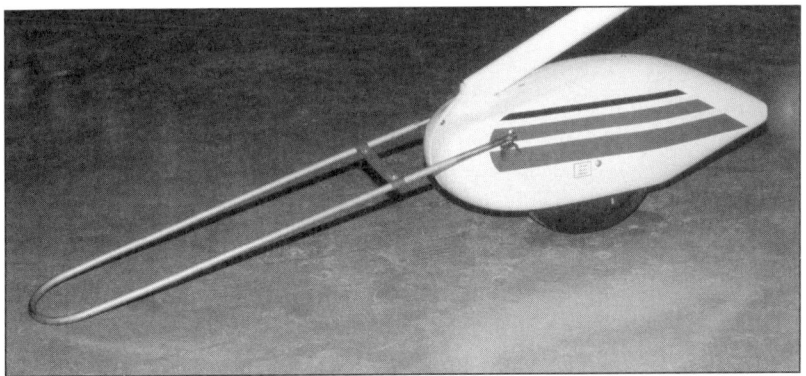

Diamond Katana DA20: A Pilot's Guide

Entering the Katana

Unlike most low-wing general aviation airplanes, the Katana is boarded from the leading edge of the wing rather than the trailing edge of the wing. A step is provided on each side of the Katana below the cockpit and ahead of the wing. A handhold is conveniently located in the forward edge of the cockpit instrument cover/glare shield. With your outside foot on the step, and your outside hand in the handhold, step up onto the cockpit floor one foot at a time, until you are standing with both feet in the airplane. Once you're standing in the Katana, place your inside hand against the seatback and lower yourself into the comfortable semi-reclined seat. Do not attempt to lower yourself into the seat with one foot outside the airplane or it will become very awkward by the time you are seated.

Once you are seated in the Katana, adjust the rudder pedal distance for your height and fasten the seat belt and shoulder harness. Lower the canopy into the closed position and slide the red latches on each side of the canopy to the forward locked position. The pilot should make sure that the canopy warning light is not illuminated. To exit the Katana, simply reverse the entry process.

Engine Starting

DA20-A1

The modern engine of the Katana DA20-A1 is easy and straightforward, the ambient conditions and the engine temperature being the principal factors to consider. As with any aircraft, always follow the manufacturer's checklist (which is in fact the reference for this text). The Rotax 912 engine presents a different starting procedure than the Continental or Lycoming engines that power other airplanes. The electric fuel pump should be turned on, the noise of the pump should be audible, and the fuel pressure warning light should not be illuminated. For the first start of the day, or if the airplane has been sitting for several hours in cool weather, the starting choke should be pulled and held on. The manual choke enables the pilot to enrich the fuel mixture when cold-starting the engine. The choke is spring-loaded to return to the off position to prevent the pilot from unintentionally operating the engine with an overly rich mixture. For cold starts, the

Section 3 Handling the Katana DA20

throttle should be in the idle position. For starts when the engine is warm, the starting choke is not used and the throttle is set 3/4-inch (2 cm) open.

After making sure that the propeller area is clear and the brakes are set, turn the ignition key to the start position. The ignition switch is in the form of a key switch, which for starting is turned past "BOTH" to the spring-loaded starter position. Cranking the engine should be limited to 10 seconds at a time, due to the danger of starter motor overheating. Normally the engine starts with in a few turns. If prolonged cranking is required, stop and investigate what the problem may be, and allow two minutes for the starter to cool down before another starting attempt is made. The starter should not be operated after the engine start, as damage to the starter will result.

The Rotax 912-powered Katana is equipped with modern altitude-compensating carburetors. There is no mixture control in the cockpit of a Katana, with the traditional idle cut-off. Just like in your automobile, turning the ignition switch to the off position shuts off the engine.

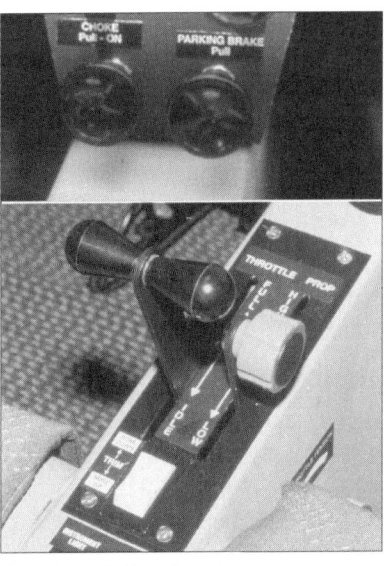

Starting choke, throttle

Ignition switch

During extreme cold weather it may be necessary to hold the choke on until the engine starts to warm up; otherwise the choke should be returned to the "OFF" position immediately after engine start. The engine instruments are located on the right-hand side of the instrument panel. Check to make sure that the oil pressure is within the green arc within 10 seconds of engine start. If the oil pressure is below 22 psi (1.5 bar) the engine should be shut down. Until the engine reaches normal operating temperature, the oil pressure may advance into the yellow arc. Readings on the voltmeter and ammeter are also checked after engine start. The warning lights should not be illuminated and the electric fuel pump should be turned off at this time. When operating the Katana in extreme low temperatures, the engine should be preheated prior to engine start. This will not only make starting easier but

Diamond Katana DA20: A Pilot's Guide

also prolongs the life of the engine. Satisfactory engine starts have been demonstrated at temperatures as low as -31°F (-35°C) outside air temperature, after a 2-hour preheat with the Tannis TAS100-27 preheat system. (*See* "Before Takeoff" checklist in Section 5.)

DA20-C1

The Continental IO-240-B engine in the DA20-C1 has fuel-injection starting, using a typical fuel-injection starting procedure. As with any aircraft engine, the ambient conditions and engine temperature are the principal factors to be considered. Extreme low temperatures require the engine to be preheated prior to engine start. As always, follow the manufacturer's checklist (which is in fact the reference used for this text).

Before starting the engine, the preflight inspection checklist should be completed, the rudder petals adjusted and locked, passenger briefing completed, and seat belts fastened. The standard items to check next are free movement of the flight controls, fuel shut-off valve open, and mixture full rich or full forward. The throttle should be set at idle, and the throttle quadrant tension device adjusted. The avionics master switch should be turned off in order to avoid damage to the aircraft radios by voltage spikes during engine start. The master switch is turned on next, which should result in illumination of the generator light. Turn on the exterior and interior lights and adjust as needed. Finally, close and secure the canopy, check that the "unlock" warning light is off. It is now time for the engine-start procedure.

Typical of a fuel-injected engine, the fuel injectors are primed prior to engaging the electric starter to turn over the engine. To prime the engine, move the throttle to the full-open or full-forward position. According to the *Airplane Flight Manual*, the electric fuel pump is turned on for 13 seconds if the engine is cold, and for only 1-3 seconds if the engine is warm. Make sure you hear an audible noise from the electric pump. The main purpose of the electric fuel pump is to prime the engine with fuel prior to engine start. This pump does not function as an emergency standby pump; it is not required to be turned on once the engine is running.

With an engine cold start, the throttle is positioned at idle or in the most aft position. If the engine has already been run and is warm, the throttle is advanced to an approximately $\frac{1}{3}$ to $\frac{1}{2}$-throttle position. The mixture should be full rich or full forward. With your feet on the toe breaks, ensuring the propeller area is clear, engage the electric starting motor. Starting cranking time should be limited to no longer then 10 seconds to prevent damage to the starter. Note that on the DA20-C1, the start

Section 3 **Handling the Katana DA20**

warning light is illuminated when the starter is engaged. Once the engine is started, release the start key and adjust the throttle to 750 rpm. Check to make sure the oil pressure is in the green arc within 30 seconds of engine start.

The DA20-C1 *Airplane Flight Manual* at the time of this writing does not address the "flooded-start" condition. For other Continental fuel-injected engines, the following procedures are recommended for this condition:

Fuel boost pump .... Off
Throttle .... Full forward
Mixture .... Idle cutoff
Ignition Key .... Start (release to BOTH with engine starts)
Mixture .... Rich
Throttle .... to Idle

Note: In the flooded-start procedure, the pilot very quickly brings the mixture from Idle Cutoff to Rich once the engine starts—and the throttle must be brought back from full forward to idle very quickly.

Taxiing

As with any aircraft, you should begin your taxi with a gradual application of engine power. As the aircraft begins to roll forward gently, apply the brakes to verify that they are working. The nose wheel on the Katana is free to castor in each direction (30° on the A1 and 60° on C1) and steering is accomplished by differential braking. The left and right brakes are independent of one another and braking action is applied in the direction of the turn. The amount of brake pressure used is determined by the desired radius of the turn: gentle taxiing turns require gentle application of brake. A tight turning radius requires more brake pressure. Taxiing the

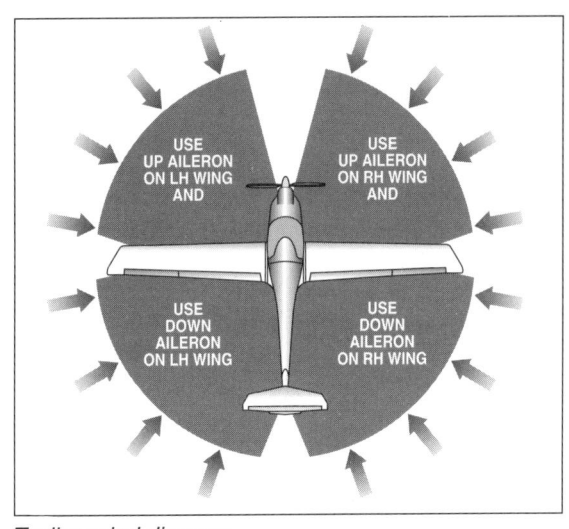

Taxiing wind diagram

3-7

Diamond Katana DA20: A Pilot's Guide

Katana is not difficult, although some practice may be needed to feel totally comfortable maneuvering the airplane on the ground.

When taxiing in the vicinity of loose sand, gravel, or standing water, minimum power should be used to avoid damaging the propeller. Be alert for obstacles, dips and bumps in the taxiway. When slowing the aircraft, the throttle should be closed first and then the brakes evenly applied. Always taxi with caution.

When taxiing with a crosswind, "opposite rudder" will be required, and up to full deflection (i.e., with a crosswind from the left, up to full right rudder) may be required as the aircraft tries to "weathervane" into the wind. Increased use of the rudder pedals and differential braking may be necessary when taxiing in crosswind conditions. The figure below shows the recommended control stick position.

(*See* "Before Taxi" checklist, Section 5.)

Power and Pre-Takeoff Checks

After taxiing to the run-up area adjacent to the active runway, the aircraft is usually positioned into the wind to aid engine cooling. Before the power check starts, the oil pressure and oil temperature should be in the green, the cylinder head temperature checked to be normal, and the fuel quantity checked with the fuel selector turned to the "ON" position.

DA20-A1

The engine is generally run up to 1,800 to 2,000 rpm (2,000 in the summer) for the propeller speed control lever to be cycled three times with a 50 to 250 rpm drop, and the ignition checked.

The Katana is equipped with dual electronic ignition systems. The systems that normally function simultaneously are checked by turning the ignition switch from "BOTH" to each independent system labeled "LEFT" and "RIGHT." When testing the individual electronic ignition systems, a small drop in rpm (no more than 150 rpm and no more than 50-rpm difference between the two) is normal and shows the ignition system is functioning normally. The throttle is reduced to 1,500 rpm and the carburetor checked by pulling the carburetor heat knob toward the pilot to the "ON" position. The rpm should show a slight decrease indicating that less dense, warmer air is entering the carburetor resulting in an enriched fuel/air mixture. Following the carburetor heat check, the throttle is reduced to idle at 950 to 1,050 rpm. Check to make sure none of the circuit breakers are popped out. Finally, one last check of engine and flight instruments, set the wing flaps to the takeoff (T/O) position, and turn the fuel pump on.

Section 3 **Handling the Katana DA20**

DA20-C1

The engine rpm is brought up to 1700 for the dual magneto check. Look for the conventional 25-150 rpm drop, with a maximum of 50 rpm difference between the two magnetos. The DA20-C1 is certified with a fixed-pitch propeller that requires no attention during the run-up. Since the engine is fuel injected, it has no carburetor heat to check. The DA20-C1 does have a vacuum gage that should be within the green arc indication during the engine run-up.

At the completion of your run-up checklist, the circuit breakers should be checked one more time, and the wing flaps set to the takeoff position.

Takeoff

DA20-A1/C1

Prior to taking the active runway, a visual check should be made to ensure the canopy is latched properly, no warning lights are illuminated, the electric fuel pump is turned on, the master switch is turned on, the ignition switch is on BOTH, carburetor heat off, wing flaps are in the T/O position, and on the DA20-A1 the propeller control lever is full forward. Upon taking the active runway and being cleared for take off, the throttle is smoothly brought full forward for maximum available takeoff power. Directional control is maintained with the rudder.

For all takeoffs, take care that your feet remain clear of the toe brakes. Keep heels on the floor of the cockpit in order to best ensure this. The memory aid "heels on the floor, brakes no more" is often used. Inadvertent pressure on the brakes can significantly slow the aircraft during the takeoff run and lead to directional control difficulties.

The Katana requires more right rudder than you might expect for a small airplane. The normal rotation speed is 51 knots for the A1, and 44 knots for the C1. The normal takeoff climb speed to clear a 50-foot obstacle published by Diamond Aircraft Industries is 60 knots for the A1, and 58 knots for C1. Best rate of climb at sea level with flaps in the takeoff position is listed as 65 knots for the A1, and 68 knots for the C1. The best angle of climb at sea level with flaps in the takeoff position is listed as 57 knots for A1, and 62 knots for C1. Control forces are light on the Katana, and care should be taken not to over-rotate during the takeoff run. After reaching a safe height, the propeller speed control lever on the A1 model should be brought back slightly to indicate 2400 rpm on the tachometer. As a general operating rule, the electric fuel pump is turned off when above 1,000 feet AGL.

(*See* "Takeoff" checklist, Section 5.)

Climbing

A cruise climb in the Katana is typically established with the propeller at 2400 rpm and the throttle at full power in the A1 and full throttle in the C1. All engine instruments should be showing normal values in the green arc. The wing flaps can be set in either the T/O position or the CRUISE position for a cruise climb. Airspeed of 65 to 75 knots works well in the Katana and provides an exceptionally good view over the nose of the airplane. Pilots should note that both the best rate-of-climb and the best angle-of-climb airspeed decreases as altitude increases.

(*See* "Climb" checklist, Section 5.)

Cruise Flight

Cruise flight is normally accomplished with a 55 to 75% power setting. Because of the airfoil design, the Katana trims for cruise in a slight nose-down attitude and will require a moment or two of attention to adjust trim as it reaches its maximum cruise speed for the selected power setting.

If turbulent conditions are encountered in flight, particular care must be taken not to exceed V_A (Design Maneuvering Speed). Maneuvering Speed for the Katana is 104 kts in the A1 and 106 knots in the C1. This is the maximum speed at which the pilot can make full or abrupt control movements without the possibility of overstressing the aircraft. The pilot should also note that the Maximum Structural Cruising Speed (V_{NO}) is 118 knots (135 mph or 218 km/h) and the Never Exceed Speed (V_{NE}), which can only be reached in a dive, is 161 knots in the A1 and 164 knots in the C1.

(*See* "Cruise" checklist, Section 5.)

Engine Handling

From the pilot's prospective, the management of the Rotax 912 in the Katana DA20-A1 is about as simple and straightforward as an automobile engine. With electronic ignition and altitude-compensating carburetors, fouled spark plugs are virtually nonexistent. As with all carburetor engines, a pilot must be aware of the possibility of carburetor icing and the use of carburetor heat. *A Pilot's Guide to the Katana* devotes an entire chapter to the topic of carburetor icing. The number one reason for aircraft engines failure is fuel exhaustion, which is mostly a matter of pilot error.

The Continental IO-240-B in the DA20-C1 is a more traditional aircraft engine, with a requirement to manually lean the engine fuel/air ratio. At engine power settings under 75% it is necessary to lean the engine with the mixture control. It should be noted that with the engine set to full throttle, it can produce less then 75% power, depending on pressure

Section 3 **Handling the Katana DA20**

altitude. Expect the engine to require leaning above 5,000 feet pressure altitude. When operating at less than 75% power and maneuvering (e.g. stalls, spins, slow flight, descents, landing approaches, landing and taxiing) the mixture should be full rich. The only exception to this procedure that is reported by Teledyne Continental Motors is when operating at very high altitudes, where the low air density may require leaning to maintain satisfactory engine operation.

Fuel management in the Katana is very simple with one fuel tank and a fuel selector that is either turned ON or turned OFF. As with all aircraft, the fuel gauge should not be trusted and the fuel quantity should be visually checked with the fuel quantity pipette. In the A1 model, the electric fuel pump should always be turned on whenever flying at and below 1,000 feet AGL. In the C1 model, you should not use the electric fuel pump once the engine is running.

Regular monitoring of the engine instruments may forewarn of an impending problem. High oil temperature, if not accompanied by a corresponding drop in oil pressure, may indicate a faulty gauge. As with most instances the action to be taken will depend on the pilot's judgment of the situation at the time. A reasonable course of action would be a diversion to a suitable airfield, while remaining alert to the possibility of a sudden engine failure.

Where a low oil pressure accompanies high oil temperature, engine failure may very well be imminent, and the pilot should act accordingly. Such a situation might occur during a prolonged slow climb in hot air temperature conditions; in this instance increasing the airspeed to provide more cooling, and reducing the power if possible, may restore oil temperature to normal. In the event of a low oil pressure reading accompanied by a normal oil temperature reading, again the gauge failure may be the culprit, and the pilot can consider actions similar to those for an oil temperature gauge failure

Aerodynamic Stalls

Note: The information in this section is not a substitute for flight instruction under the guidance of a flight instructor familiar with the aircraft and its characteristics.

The Katana DA20 exhibits conventional and relatively benign high angle of attack and stall behavior. The Wortmann FX 63-137/20 HOAC airfoil is optimized for low speed handling qualities. In addition to the wing itself, the turned-up wing tips, called "vortips," reduce drag, improve roll stability, and enhance aileron effectiveness at low airspeed. It is, however, somewhat difficult to achieve a full stall condition. The trade-off is that the modern aerodynamic features of the Katana make for a very safe and confidence-building airplane for low-time beginning pilots. This is an important quality for a primary flight training aircraft.

3-11

The stall warning horn activates 5 to 10 knots above the stall airspeed. The airspeed indicator is unreliable near the stall airspeeds and tends to under-read considerably. The use of power will lower the stalling speed, while turning flight raises the stalling speed. The Katana has a very definite stall buffet that becomes even more pronounced as flaps are deployed from CRUISE, to TAKEOFF position (15°), to LANDING flaps (40°). The wing-tip design actually creates a flow of air over the aileron to improve roll control during flight at slower airspeeds. In coordinated flight with full aft stick the Katana still provides aileron roll authority. The Katana provides significant warning of an impending stall and very docile flight characteristics when the wing is aerodynamically stalled.

The entry altitude for aerodynamic stall practice should be high enough so that recovery from the maneuver will occur at no less than 1,500 feet above ground level (AGL). An entry altitude of at least 2,500 feet AGL is suggested to allow sufficient altitude to practice descents and glides. All stall practice should be proceeded by visual clearing turns, either a 360°-turn, a 180°-turn, or two 90°-turns. It is important to continue to scan for aircraft traffic during the maneuver.

You should learn to use sight, feel, and sound to recognize an imminent stall condition.

Sight: attitude of the airplane in the stall and the airspeed indicator. Note that vision would be of little use to detect a stall that occurs in a normal attitude. For more information consult a flight training handbook, flight maneuvers manual, or the FAA *Flight Training Handbook*.

Feel: the controls are mushy and not as effective as at higher airspeeds and lower angles of attack. The stall buffet may be felt as the stall approaches.

Sound: tones and intensity of sounds incident to flight decrease as airspeed decreases. When the stall is almost complete, vibration and its incident noises often increase greatly. While stall warning devices are an important safety feature, and the Katana's stall warning system warns you 5 to 10 knots before the stall occurs, you must also learn to recognize stalls without this aid.

All pilots should periodically practice stalls, power-on and power-off, without flaps, with partial flaps, and with full flaps, straight ahead and with turns. Again, always practice stalls at an altitude that will allow for recovery to be made no lower than 1,500 feet AGL. All pilots should become proficient at these maneuvers so they can recognize a stall before it happens, and so that if they enter an inadvertent stall, they will have the skill necessary to recover with minimum loss of altitude. The typical height loss for a full stall with conventional recovery (using power) is about 200 feet.

Section 3 **Handling the Katana DA20**

Spins

Note: The information in this section is not a substitute for flight instruction under the guidance of a flight instructor familiar with the aircraft and its characteristics.

The Katana DA20 is approved for intentional spins with the flaps up and makes an excellent platform for spin training. As you are getting set up to experience spin entry and recovery, a few items should be addressed. All loose items should be stowed and the safety harnesses firmly secured. **Even though the FAA minimum recovery altitude is 1,500 feet above ground, it is strongly recommended that you plan an entry altitude high enough that your recovery will occur no less than 2,000 feet above ground level.** A higher entry altitude will assist in maintaining pilot orientation by providing a better field of view.

In the Katana DA20, plan for at least 1,000 feet of altitude loss for a one-turn spin entry and recovery, and allow an additional 300–500 feet of altitude loss for each additional turn. Limiting yourself to two-turn spins until you are completely familiar with the characteristics of the airplane decreases the possibility of losing orientation. *A helpful hint:* do spins high over a road to make it easier to count the number of turns in the spin. With these considerations in mind, the last remaining task is to make sure the area in which you are doing your spins is clear of other aircraft traffic. It is difficult to see and avoid other airplanes once you are looking at the earth from a spin.

The first thing that you need to think about prior to entering a spin is the configuration of the airplane. Note that in the Katana DA20, there is a placard directly in front of the pilot that states spins are permitted only with flaps up. One reason for this is that when the flaps are extended, the rate of turn in the spin increases dramatically to the point that you are spinning so fast, it is difficult to remain oriented. Verify that the flaps are in the "UP" or "CRUISE" position. The electric fuel pump should be turned on, the throttle at idle, carburetor heat pulled on, and the propeller control moved full forward for high rpm.

The Katana should be trimmed for a spin entry speed of 65 knots in the A1 model, and 58 knots in the C1. This speed should then be reduced with elevator back pressure at the rate of 2–3 knots per second. When the stall warning horn sounds, apply full aft stick and full rudder in the direction of the desired rotation. *A note of caution:* if a good clean stall break is not achieved, the spin entry may develop into a spiral dive. A spiral dive is a steep, diving turn with increasing airspeed. (This increasing airspeed can be difficult to overcome prior to exceeding V_{FE}, which supports the fact that spins should only be practiced with flaps in the up position when flying a Katana DA20.) Conventional control input will recover the airplane since the wing is not in a stalled condition. It should also be noted that spins in an airplane with an aft center of gravity may oscillate in yaw and pitch attitude. This condition has no effect on recovery procedure or recovery time.

The first step in recovery from a spin is to make sure that the throttle is pulled back to idle power. Apply rudder in the opposite direction of the spin. Almost simultaneously, ease the control stick forward until the spinning stops. As soon as the rotation stops, the rudder should be neutralized. Check to make sure that the wing flaps are in the UP or CRUISE position. The airplane is now out of the spin and in a dive. Ease the stick back to pull out of the dive, being careful not to overstress the airplane or exceed the maximum permissible airspeed.

The first turn or two of a spin is typically considered the "incipient phase" and is the transition between the stall and a fully developed spin. In a fully developed spin, there is a balancing of aerodynamic and inertial forces, producing what is called "autorotation," a repeatable pattern of rotation and pitch attitudes. Recovery occurs when autorotation has stopped. You may think you are looking straight down at the earth during the spin. However, your actual attitude is probably closer to 45 to 60 degrees nose-down attitude. Expect to pull 2.5 to 3 Gs during the pullout.

Descent

The descent may be powered or glide. In the DA20-A1, the propeller rpm is typically set at 1900–2400 rpm. The manifold pressure is set as required, with carburetor heat applied when power is brought below 20 inches manifold pressure. In the DA20-C1 with a fixed-pitch propeller, the rpm must be kept in the green arc on the tachometer, and the cylinder head temperature should not fall below 300°F for more than five minutes and never less than 240°F. Carrying power on the descent will provide for a higher airspeed (and ground speed) descent, and will keep the engine at a warmer temperature than in a total power-off glide. Always keep in mind the airspeed limitations of the aircraft, maneuvering speed, and maximum structural cruising speed. Typical of aerodynamically clean, high-performance airplanes, the Katana picks up speed quickly in a power-on descent.

There is no prohibition against slipping the Katana. However, the small fuselage of the Katana does not produce much drag in a side slip, and thus has a smaller effect on the descent rate than would be expected of other airplanes. For a relatively slow and very steep descent in a Katana, try full flaps and up to an 78-knot airspeed (do not exceed 81 knots in the A1 and 78 knots in C1). If 78 knots is carried through to the approach, the airspeed will bleed off quickly because the lightweight (low mass) airframe

Section 3 **Handling the Katana DA20**

does not develop much inertia to slow down. This technique works very well for short approaches as an advanced pilot technique that should be introduced to the new pilot by a qualified instructor.

(*See* "Descent" checklist, Section 5.)

Landing

The Katana DA20 is a relatively simple, fixed-landing-gear, single-fuel-tank airplane with a correspondingly simple pre-landing checklist. Prior to entering the airport traffic pattern and at and below 1,000 feet AGL, check the following items: seat belts—fastened and properly adjusted, and electric fuel pump—turned on. Entering the downwind leg of the airport traffic pattern it is suggested that you reduce the power and slow the Katana to 80 knots.

The following is the typical landing profile for an ideal airport traffic pattern and landing. At most airports these days it is not possible to fly "the ideal traffic pattern" because of the volume of traffic that must be dealt with. It is very common for aircraft to have to extend the downwind leg, or be asked to make a short approach from the downwind leg, The description presented is for the purposes of orientation to the flight considerations of the Katana DA20-A1.

Downwind abeam the touchdown numbers, manifold pressure is reduced to 15 inches in the DA20-A1 or 1500 rpm in the DA20-C1, carburetor heat pulled on, and wing flaps selected to the takeoff position. This power setting and configuration should provide a descent rate of about 200 feet per minute. When the aircraft is approximately 45° past the runway touchdown numbers, the base leg turn is initiated. On base leg, the manifold pressure is reduced to 12 inches on the DA20-A1 or 1200 rpm in the DA20-C1 and the airspeed slowed to 70 knots, with a resulting descent rate of 300 feet. The turn to final is made to intercept the extended centerline of the runway. Established on final approach, landing flaps (40°) are selected, the propeller speed control lever is moved full forward, and the pitch and power is set for a 57 to 60-knot final approach airspeed in a DA20-A1, and 52 to 55 knots in a DA20-C1. Past the landing threshold and at the appropriate height over the runway, the landing flare is commenced. The throttle is retarded to idle and the nose of the airplane is gently raised for the landing flare and touchdown on the main wheels.

The Katana is a very honest, straightforward airplane to fly and land. The control stick uses push/pull rods directly connecting to the elevator and aileron. The control forces are light and positive in feel, and as a result of the direct connection to the control surfaces the pilot can actually feel the aerodynamic forces bleeding off as the airspeed decays. While the

Diamond Katana DA20: A Pilot's Guide

Katana is an easy airplane to fly, it must be understood that there is no substitute for flight instruction in the proper technique with a qualified flight instructor.

After Landing

After landing, and once you have taxied clear of the active runway, bring the wing flaps up to the cruise position, push carburetor heat off (in the DA20-A1), turn off the electric fuel pump, and turn the transponder to standby. After taxiing to parking, bring the throttle to idle and consult the engine shutdown checklist. Check the ELT by listening to 121.5 MHz. Turn off all electricals; turn off the avionics master switch, followed by the ignition switch and then the master switch.

(*See* "After Landing and Clearing the Runway" checklist, Section 5.)

Parking and Tie Down

The aircraft is generally parked into the wind. It is a good practice to stop with the nose wheel straight so that the rudder is not deflected. At this point you should verify that all switches are off. Next, the rudder pedal assembly on the pilot's side is pulled to the full forward position and the control lock installed (as per the instructions on the control lock). The control lock works best when the control stick is trimmed nose down, which may require that the master

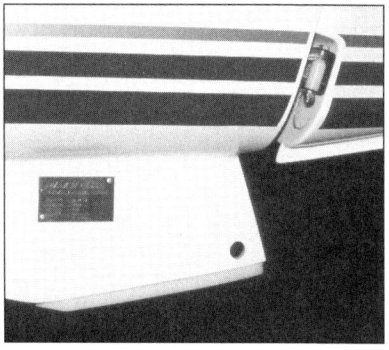

Mooring or tie-down hole

switch be turned on for a few moments to activate the trim switch. As a standard safety precaution, whenever the aircraft master switch is turned on, the propeller area should be clear. The flagged pitot static cover is placed over the pitot static probe and the stall-warning plug fitted into the stall warning opening on the left wing. One or more of the aircraft's wheels should be chocked as good operational practice.

If the airplane is going to be unattended for an extended period of time or if windy conditions exist, the aircraft should be secured to the ground anchor points with mooring straps, ropes or chains. Mooring rings are provided near the midpoint of each wing. The ventral fin of the Katana has a hole in it for tying the tail of the airplane to the ground.

Section 4
Carburetor Icing Supplement

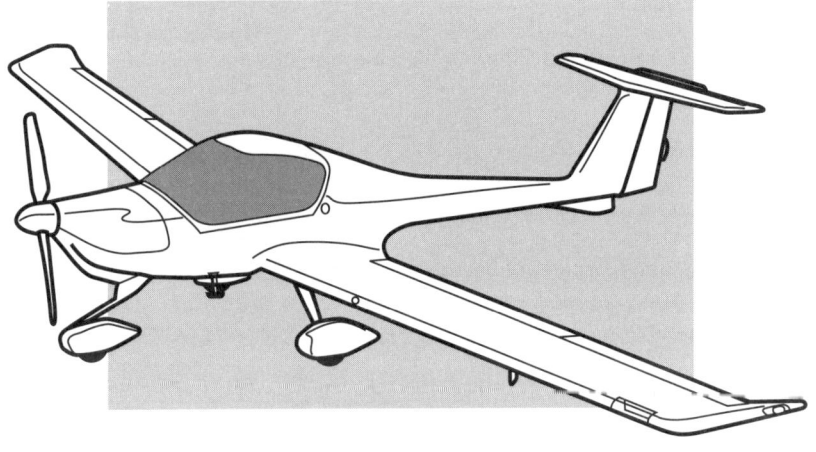

Diamond Katana DA20: A Pilot's Guide

Section 4 Carburetor Icing Supplement

Introduction

The Rotax 912 engine in the DA20-A1 Katana uses a carburetor as a fuel metering system, and as such is potentially subject to the threat of carburetor icing just as any other carbureted reciprocating engine. The carburetor icing discussion that follows is generic to all engine types, and is provided to develop an awareness and understanding of carburetor icing.

Carburetor Icing

Almost certainly the most common cause of engine rough-running, and complete engine failure, is carburetor icing. Despite this, carburetor icing remains a widely misunderstood subject. Many pilots' knowledge of the subject is limited to a feeling that the carburetor heat should be used regularly in flight, without really knowing the symptoms of carburetor icing or the conditions most likely to cause its formation.

How Carburetor Icing Forms

IMPACT ICING occurs when ice forms over the external air inlet (air filter) and inside the induction system leading to the carburetor. This type of icing occurs with the temperature below 0°C while flying in clouds, or in precipitation (i.e., rain, sleet or snow). These conditions are also conducive to airframe icing, and the aircraft is not cleared for flight into known icing conditions, which clearly these are. So, assuming the aircraft is operated legally within its limitations, this form of icing should not occur, and is not considered further.

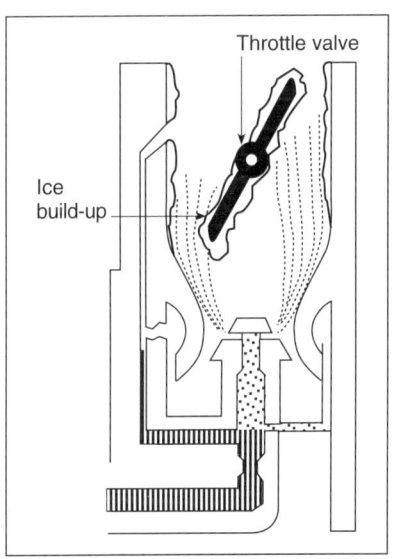

CARBURETOR ICING is caused by a temperature drop inside the carburetor, which can happen even in conditions where other forms of icing will not occur. The causes of this temperature drop are twofold:

1. FUEL ICING—the evaporation of fuel inside the carburetor. Liquid fuel changes to fuel vapor and mixes with the induction air, causing a large temperature drop. If the temperature inside the carburetor falls below 0°C, water vapor in the atmosphere condenses into ice, usually on the

walls of the carburetor passage adjacent to the fuel jet and on the throttle valve. Generally, fuel icing is responsible for around 70% of the temperature drop in the carburetor.

2. THROTTLE ICING—the temperature loss caused by the acceleration of air and consequent pressure drop around the throttle valve. This effect may again take
the temperature below 0°C, and water vapor in the inlet air will condense into ice on the throttle valve. This practical effect is a demonstration of Bernoulli's Principle.

As fuel and throttle icing generally occur together, they are known just as carburetor icing.

Conditions Likely to Lead to Carburetor Icing

Two criteria govern the likelihood of carburetor icing conditions: the AIR TEMPERATURE and the RELATIVE HUMIDITY.

The ambient air temperature is important, but not because the temperature needs to be below 0°C, or even close to freezing. The temperature drop in the carburetor can be up to 30°C, so carburetor icing can (and does) occur in hot ambient conditions. It is no wonder carburetor icing is sometimes referred to as refrigeration icing. Carburetor icing is considered a possibility within the temperature range of -10°C to +30°C.

The relative humidity (a measure of the water content of the atmosphere) is a major factor. The greater the water content in the atmosphere (the higher the relative humidity), the greater the risk of carburetor icing. That said, the relative humidity (RH) does not to have to be 100% (visible water droplets—cloud, rain), for carburetor icing to occur. Carburetor icing is considered a possibility at relative humidity values as low as 30%. Herein lies perhaps the real danger of carburetor icing, that it can occur in such a wide range of conditions. Obviously the pilot must be alert to the possibility of carburetor icing at just about all times. Flight in or near clouds, or in other visible moisture (i.e., rain) might be an obvious cause of carburetor icing, but—*visible moisture does not need to be present for carburetor icing to occur.*

Section 4 **Carburetor Icing Supplement**

Carburetor Icing Conditions

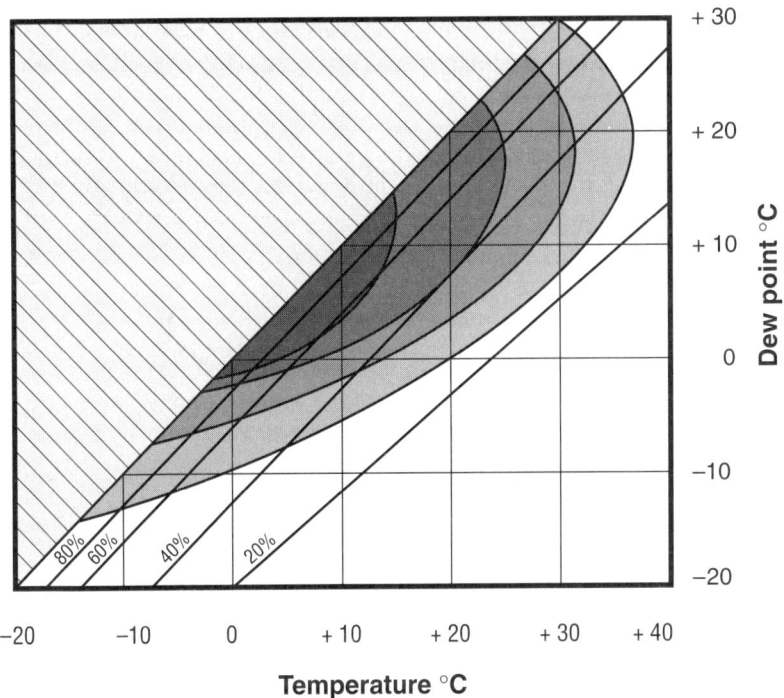

- 100% Relative humidity
- Serious icing – any power
- Moderate icing – cruise power
  Serious icing – descent power
- Serious icing – descent power
- Light icing – cruise or descent power

Diamond Katana DA20: A Pilot's Guide

Symptoms of Carburetor Icing

In the DA20-A1, fitted with a constant-speed propeller, the symptoms of carburetor icing are straightforward. A loss of manifold pressure will be the first symptom, although this is often first noticed as a loss of altitude. As the icing becomes more serious, engine rough-running may occur.

Carburetor icing is often detected during the use of the carburetor heat. Normally when the carburetor heat is used, a small drop in rpm occurs; when the control is returned to cold (OFF), the rpm restores to the same as before the use of carburetor heat. If the rpm restores to a figure higher than before the carburetor heat was used, it can be assumed that some form of carburetor icing was present.

Use of Carburetor Heat

Apart from the normal check of carburetor heat during the power checks, it may be necessary to use the carburetor heat on the ground if carburetor icing is suspected. Safety considerations apart, the use of carburetor heat on the ground should be kept to a minimum, because the hot air inlet is unfiltered and sand or dust can enter the engine, increasing engine wear.

Carburetor icing is generally considered to be very unlikely with the engine operating at above 75% power, that is, during the takeoff and climb. Carburetor heat should not be used with the engine operating at above 75% power (i.e., full throttle) as detonation may occur. Detonation is the uncontrolled burning of fuel in the cylinders, literally an explosion, and will cause serious damage to the engine very quickly. Apart from the danger of detonation, the use of carburetor heat reduces the power the engine produces. In any situation where full power is required (takeoff, climb, go-around) the carburetor heat must be off (cold).

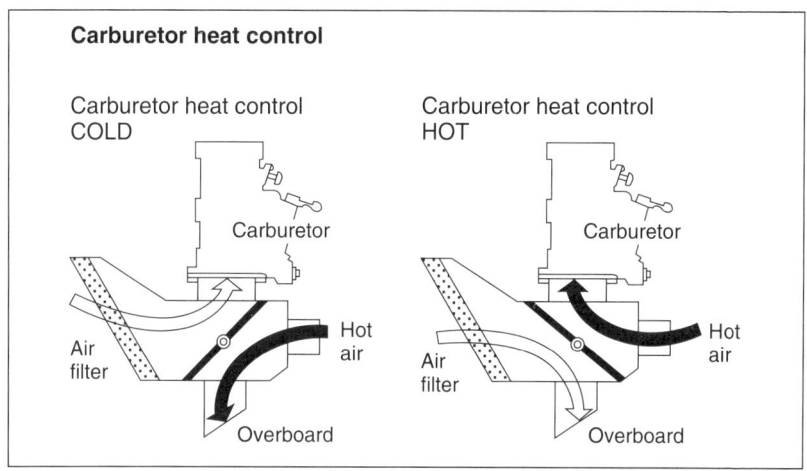

4-6

Section 4 **Carburetor Icing Supplement**

Very few operators recommend the use of anything other than FULL carburetor heat. A normal carburetor icing check will involve leaving the carburetor heat on (hot) for 5 to 10 seconds, although the pilot may wish to vary this dependent on the conditions. The use of carburetor heat does increase the fuel consumption, and this may be a factor to consider if the aircraft is being flown towards the limit of its range/endurance in possible carburetor icing conditions.

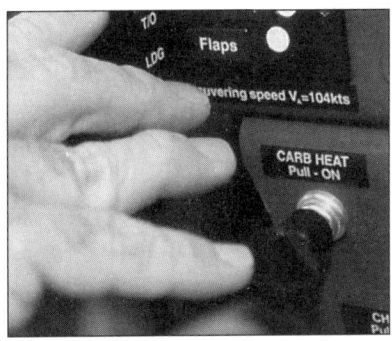

Cockpit carburetor heat control

With carburetor icing present, the use of carburetor heat may lead to a large drop in rpm, with rough running. The instinctive reaction is to put the carburetor heat back to cold (OFF), and quickly—this is, however, the wrong action. Chances are this rough running is a good thing, and the carburetor heat should be left ON (hot) until the rough running clears, and the rpm rises. In this instance, the use of carburetor heat has melted a large amount of accumulated icing, and the melted ice is passing through the engine causing temporary rough running.

Care should be taken when flying in very cold ambient conditions (below -10°C). In these conditions, the use of carburetor heat may actually raise the temperature in the carburetor to that most conducive to carburetor icing. Generally, when the temperature in the carburetor is below -8°C, moisture forms directly into ice crystals which pass through the engine.

The rpm loss normally associated with the use of carburetor heat is caused by the reduced density of the hot air entering the carburetor, leading to an over-rich mixture entering the engine. If the carburetor heat has to be left constantly ON (hot)—such as flight in heavy rain and clouds—it may be advisable to lean the mixture in order to maintain rpm and smooth engine running.

It is during the descent (and particularly the glide descent) that carburetor icing is most likely to occur. The position of the throttle valve (almost closed) is a contributory factor, and even though the carburetor heat is normally applied throughout a glide descent, the low engine power will reduce the temperature of the hot air selected with the carburetor heat control. In addition, a loss of power may not be readily noticed, as the propeller is likely to windmill even after a complete loss of power. A full loss of power may only be apparent when the throttle is opened at the bottom of the descent. This is one good reason for opening the throttle to "clear the engine" at intervals during a glide descent.

Diamond Katana DA20: A Pilot's Guide

Section 5
Katana DA20 Checklists

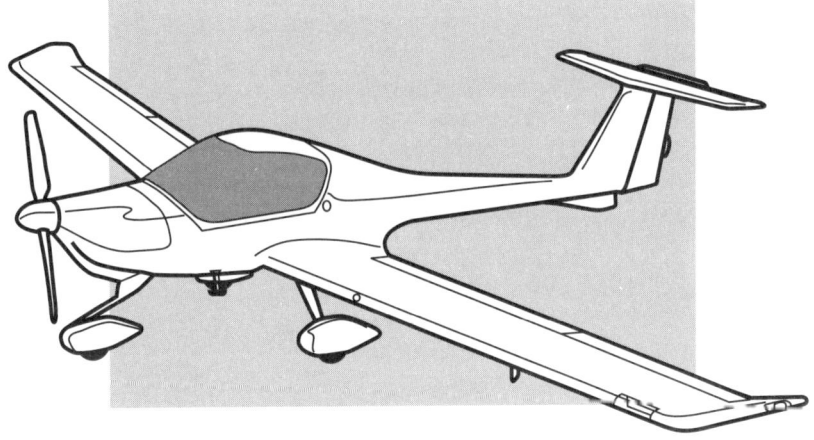

Diamond Katana DA20: A Pilot's Guide

Section 5 **Katana DA20 Checklists**

The official approved *Airplane Flight Manual* for the operation of the airplane is always considered the definitive aircraft checklist. This *Pilot's Guide* is a general information guide to the Katana, and the following checklists are provided to be consistent with that purpose. This guide is in no way designed to substitute or replace the official checklist for the Katana DA20.

Approaching Aircraft

Note: "*Visual inspection*" *(as listed below) is defined as checking for defects, cracks, delaminations, excessive play, insecure or improper mounting, and general condition. Additionally, visual inspection of control surfaces includes checking for freedom of movement.*

1. Check for and remove any tie-downs, external control locks, pitot cover, stall warning plug and wheel chocks.
2. Look for oil and fuel spillage from aircraft.
3. Remove ice and frost from all surfaces.
4. Check for access to taxiways, obstructions, loose gravel, etc.
5. Look to see if aircraft is on a level surface. Sloping ground will effect the visual check of fuel quantity.

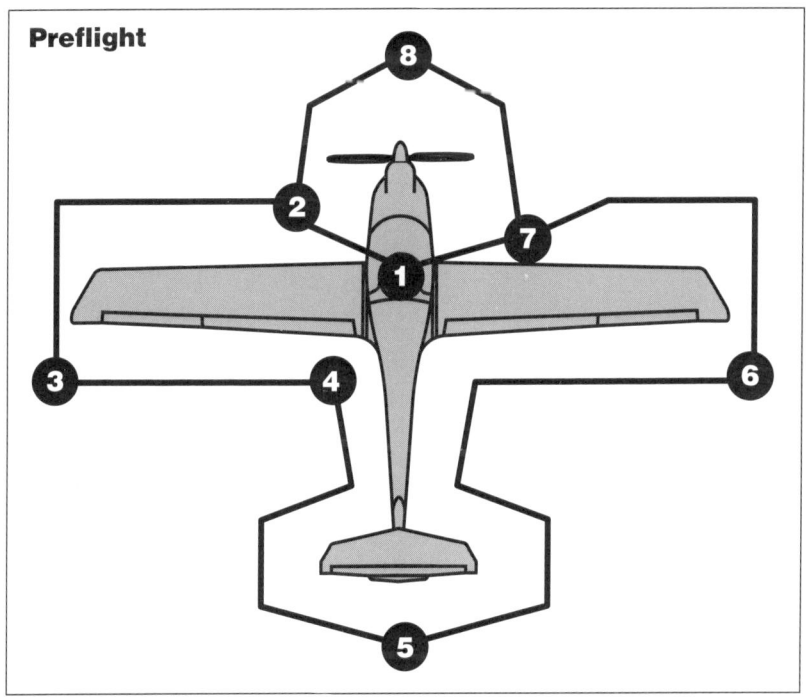

Preflight

Diamond Katana DA20: A Pilot's Guide

Cockpit ❶

1. Structural temperature indicator .... less than 55°C (131°F)
2. Aircraft documents .... check
3. Flight control lock .... remove
4. Flight controls .... free and correct
5. Ignition key .... pulled out
6. Carburetor heat (A1 only) .... free, OFF
7. Cabin heat .... free
8. Choke (A1 only) .... free, self-reseating
9. Parking brake .... free
10. Throttle .... free, idle
11. Mixture control (C1 only) .... free, idle cut-off
12. Propeller speed control (A1 only) .... free, maximum rpm
13. All switches .... OFF
14. Master switch (battery) .... ON
15. Warning lights (generator, fuel pressure, & canopy) .... illuminated
16. Fuel quantity .... sufficient
17. Engine gauges, ammeter, and voltmeter .... check
18. Circuit breakers .... check in
19. Map light .... operational
20. Instrument lights .... operational, dimmable
21. Elevator trim .... check movement, set neutral
22. Wing flaps .... check, extend, and retract fully
23. Trim and flap indicator lights .... operational, dimmable
24. Exterior lights .... operational as required
25. Master switch .... OFF
26. Foreign object inspection .... complete
27. Emergency locator transmitter (ELT) .... EBC Model 502
28. Fire extinguisher .... EBC Model 102A, OFF
29. Baggage .... stow, baggage net attached
30. Canopy .... clean, undamaged

Section 5 **Katana DA20 Checklists**

Left Main Landing Gear ❷
1. Landing gear strut .... visual inspection
2. Wheel fairing .... visual inspection
3. Tire pressure (33 psi / 2.3 bar) .... check
4. Tire, wheel, brake .... visual inspection
5. Wheel chocks .... remove

Left Wing ❸
1. Entire wing .... visual inspection
2. Stall warning inlet .... remove plug, check unobstructed
3. Pitot-static probe .... remove cover, check unobstructed
4. Tie down .... removed
5. Taxi and landing light .... visual inspection
6. Wing tip, position lights, and strobe .... visual inspection
7. Aileron balancing weight .... visual inspection
8. Aileron including inspection panel .... visual inspection
9. Wing flap including inspection panel .... visual inspection

Fuselage ❹
1. Skin .... visual inspection
2. Fuel tank vent .... unobstructed
3. Fuel drains .... drain and check
4. Fuel quantity .... use fuel dipstick for visual inspection
5. Fuel cap .... secure
6. Antennas .... visual inspection

Empennage ❺
1. Stabilizers and control surfaces .... visual inspection
2. Tie down .... remove
3. Trim tab .... visual inspection

Diamond Katana DA20: A Pilot's Guide

Right Wing ❻
1. Entire wing .... visual inspection
2. Wing flap .... visual inspection
3. Aileron including inspection panel .... visual inspection
4. Aileron balancing weight .... visual inspection
5. Wing tip position lights and strobe .... visual inspection
6. Tie down .... remove

Right Main Landing Gear ❼
1. Landing gear strut .... visual inspection
2. Wheel fairing .... visual inspection
3. Tire pressure (33 psi / 2.3 bar) .... check
4. Tire, wheel, brake .... visual inspection
5. Wheel chocks .... remove

Nose ❽
1. Oil quantity .... check level, refill as required
2. Coolant quantity (A1 only) .... check level, refill as required
3. Cowling .... visual inspection, no fluids leaking
4. Air intakes .... unobstructed
5. Propeller .... visual inspection, ground
6. Propeller blades (A1 only) .... perform pitch check by hand
7. Spinner .... visual inspection
8. Nose gear .... visual inspection, towbar removed
9. Wheel fairing .... visual inspection
10. Tire pressure (26 psi/1.8 bar) .... check
11. Tire and wheel .... visual inspection
12. Wheel chocks .... remove
13. Engine compartment drain .... unobstructed
14. Exhaust stack (caution—may be HOT!) secure and in good condition

Section 5 **Katana DA20 Checklists**

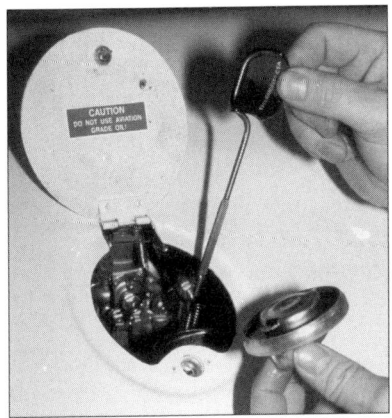

Checking oil

Before Starting Engine

1. Preflight inspection .... complete
2. Rudder pedals .... adjust
3. Passenger briefing .... perform
4. Safety harnesses .... fastened and properly adjusted
5. Parking brake .... set
6. Flight controls .... free and correct
7. Fuel shut-off valve .... open
8. Mixture (C1 only) .... full rich
9. Carburetor heat (A1 only) ....free, OFF
10. Propeller speed control lever (A1 only) .... free, maximum rpm
11. Friction device on throttle quadrant .... adjust
12. Avionics master switch .... OFF
13. Master switch (battery/generator) .... ON
14. Generator warning light .... illuminated
15. Fuel pressure warning light .... illuminated
16. Canopy locking warning light .... illuminated
17. Exterior lights .... as required
18. Instrument panel lights .... as required
19. Canopy .... close and lock
20. Canopy warning light .... OFF

5-7

Diamond Katana DA20: A Pilot's Guide

Because of different starting procedures for the Rotax and the Continental powered aircraft, the "Starting Engine" checklist is presented for both the DA20-A1 and the DA20-C1.

Starting Engine
DA20-A1 (Rotax 912)
Note: Under certain circumstances, activation of the fuel pressure warning light might take as long as 10 minutes after shutting down engine or switching off the electric fuel pump.

1. Electric fuel pump .... ON (sound of pump audible)
2. Fuel pressure warning light .... OFF
3. Throttle Cold start .... idle
            Warm engine .... open 3/4 inch (2 cm)
4. Choke Cold start .... ON, fully pulled out
          Warm engine .... OFF
5. Toe brakes .... hold
6. Propeller area .... clear
7. Ignition key .... start
8. Choke .... after start, push OFF
9. Throttle .... 1,500 rpm maximum
10. Oil pressure .... green, within maximum of 10 seconds
11. Generator warning light .... OFF
12. Exterior lights .... as required
13. Electric fuel pump .... OFF

DA20-C1 (Continental IO-240-B)
1. Throttle .... full forward (open) for priming
2. Mixture .... full rich
3. Electric fuel pump   Engine cold .... ON 3 seconds
                                                     (sound of pump audible)
                         Engine warm .... ON 1–3 seconds
                                                           (leave pump on)
4. Throttle   Cold start .... 1/4 inch open
                        Warm start .... approx. 1/2 inch open
5. Toe breaks .... hold
6. Propeller area .... clear ("Clear Prop")

Section 5 Katana DA20 Checklists

7. Ignition key .... start
8. Start warning light .... off when key is released
9. Throttle .... 800–1,000 rpm
10. Oil pressure .... indication after maximum of 30 seconds
11. Generator warning light .... OFF
12. Exterior lights .... as required
13. Electric fuel pump .... OFF

Before Taxiing
1. Avionics master switch .... ON
2. Flight instruments and avionics .... set
3. Engine gauges .... check
4. Voltmeter .... verify in the green
5. Warning lights (generator, fuel pressure, canopy) .... push to test
6. Parking brake .... released

Taxiing
1. Brakes .... check
2. Directional control .... check
3. Flight instruments/avionics .... check
4. Compass .... check

Because of different engine runup procedures for the Rotax and Continental powered aircraft, the "Before Takeoff Checklist" is presented for both the DA20-A1 and DA20-C1 aircraft.

Before Takeoff (Engine Runup)
DA20-A1 (Rotax 912)
1. Nose wheel .... straight
2. Toe brakes .... hold
3. Safety belts and harnesses .... fastened and properly adjusted
4. Canopy .... closed and locked
5. Fuel pressure warning light .... OFF
6. Electric fuel pump .... ON

5-9

Diamond Katana DA20: A Pilot's Guide

7. Fuel shut-off valve .... check open
8. Fuel quantity indicator .... check, sufficient
9. Engine instruments .... check within green range
10. Trim .... neutral
11. Flight controls .... free and correct
12. Throttle .... 1,800–2,000 rpm
13. Propeller control .... cycle 3 times (200–300 rpm drop)
14. Ignition switch .... cycle L-BOTH–R-BOTH
    (maximum rpm drop 150, 50 difference R and L)
15. Throttle .... 1500 rpm
16. Carburetor heat .... ON (maximum 50 rpm drop)
17. Throttle .... idle
18. Carburetor heat .... OFF
19. Radios .... ON and tuned
20. Transponder .... altitude
21. Lights .... as required
22. Clearance .... obtained (as required)
23. Before taking the runway .... visually check clear
24. Wing flaps .... takeoff
25. Parking brake .... release

DA20-C1 (Continental IO-240-B)

1. Nose wheel... straight
2. Toe brakes .... hold
3. Safety belts and harnesses .... fastened and properly adjusted
4. Canopy .... closed and locked
5. Fuel pressure warning light .... OFF
6. Fuel shut-off valve .... check open
7. Fuel quantity indicator .... check, sufficient
8. Engine instruments .... check within green range
9. Trim .... neutral
10. Flight controls .... free and correct
11. Throttle .... 1700 rpm

Section 5 **Katana DA20 Checklists**

12. Ignition switch .... cycle L-BOTH–R-BOTH
    (maximum rpm drop 150, 50 difference R and L)
13. Mixture vacuum gage .... full rich
14. Vacuum gage .... check within green range
15. Throttle .... idle
16. Circuit Breakers .... check
17. Radios .... ON and tuned
18. Transponder .... squawk altitude
19. Lights .... as required
20. Clearance .... obtained (as required)
21. Before taking the runway .... visually check clear
22. Wing flaps .... takeoff
23. Parking brake .... release

Takeoff
1. Electric fuel pump   (A1) .... check ON
                         (C1) .... check OFF
2. Master switch (battery/generator) .... check ON
3. Ignition switch ... .,,,,,,,,... check BOTH
4. Carburetor heat (A1 only) .... check OFF
5. Wing flaps .... check takeoff
6. Propeller speed control (A1 only) .... maximum rpm
7. Throttle   (A1) .... full, 2400 to 2500 rpm
                         (C1) .... full, min 2000 rpm
8. Elevator—at beginning of roll .... neutral
9. Directional control .... maintain with rudder

Normal Takeoff

Lift nose wheel	(A1)	....51 kts / 95 km/h
	(C1)	....44 kts / 81 km/h
Climb speed	(A1)	....65 kts / 120 km/h
	(C1)	....58 kts / 107 km/h

5-11

Short-Field Takeoff (over a 50-foot obstacle)

Lift nose wheel	(A1)	57 kts / 106 km/h
	(C1)	52 kts / 96 km/h
Climb speed	(A1)	60 kts / 110 km/h
	(C1)	58 kts / 107 km/h

Soft-Field Takeoff

Pre-takeoff checks and runup .... Perform over hard surface if possible

Control stick .... Full aft, keeping weight off the nose wheel

Throttle .... Carry enough power to keep the aircraft moving.

During takeoff roll:

Elevator .... Use enough aft stick to keep the aircraft's weight on the main wheels.

After lift off .... Immediately lower the nose to accelerate to 60kts (110 km/h) while in ground effect.

Climb airspeed (A1) .... 60 kts / 110 km/h Best angle of climb;
(C1) .... 62 kts / 115 km/h Best angle of climb
(A1) .... 65 kts / 120 km/h Best rate of climb
(C1) .... 68 kts / 126 km/h Best rate of climb

Climb

1. Mixture (C1) .... full rich
2. Propeller speed control lever (A1 only) .... 2400 rpm
3. Throttle .... full
4. Engine gauges .... check within green range
5. Wing flaps .... takeoff or cruise
6. Climb speed (V_Y)   (A1) .... 65 kts / 120 km/h
(C1) .... 75 kts
7. Electric fuel pump (A1) .... OFF, upon reaching a safe altitude, typically 1,000 feet AGL
8. Fuel pressure warning light .... monitor during flight
9. Elevator trim .... adjust

Note: The best rate of climb decreases with increasing altitude.

Section 5 **Katana DA20 Checklists**

Cruise
1. Throttle .... as required
2. Mixture (C1 only) .... Lean to 25° rich of peak EGT.
   Do not lean by EGT above 75% power
3. Propeller speed control lever (A1 only) .... 1900–2400 rpm
4. Wing flaps .... cruise
5. Trim .... as required
6. Engine gauges .... check, monitor during cruise

Descent
1. Flight instruments and avionics .... adjust
2. Mixture (C1 only) .... full rich
2. Throttle .... as required
3. Propeller (A1 only) .... 1900–2400 rpm
4. Carburetor heat (A1 only) .... as required

Accelerated/Fast Descent
1. Propeller   (A1) .... 2400 rpm
                     (C1 .... 2800 rpm
2. Throttle .... Idle
3. Carburetor heat (A1 only) .... ON
4. Wing flaps .... CRUISE (Up)
5. Airspeed .... 118 kts / 135 mph / 218 km/h

Landing Approach
1. Seat belts and harnesses .... fastened and properly adjusted
2. Electric fuel pump (A1 only) .... ON
3. Lights .... as required
4. Master switch (battery/generator) .... check ON
5. Ignition switch .... check BOTH
6. Carburetor heat (A1 only) .... ON
7. Throttle .... as required
8. Airspeed   (A1) .... maximum 81 kts / 93 mph / 150 km/h
                    (C1) .... maximum 78 kts / 144 km/h

5-13

9. Wing flaps .... takeoff
10. Trim .... as required
11. Propeller speed control (A1 only) .... maximum rpm
12. Wing flaps .... landing
13. Approach speed   (A1) .... 57 kts / 66 mph / 106 km/h
                         (C1) .... 52 kts / 96 km/h

Soft-Field Landing

1. Flaps .... full / landing flaps
2. Final approach speed   (A1) .... 57 kts / 106 km/h
                              (C1) .... 52 kts / 96 km/h
3. Throttle .... As required, to touch down on main wheels in a nose-high attitude. Continue to hold the aft and carry enough power to keep weight off the nose wheel, until the aircraft is on solid ground.

Short-Field Landing

1. Flaps .... full / landing flaps
2. Final approach .... steeper than normal, allowing for obstacle clearance
3. Final approach speed   (A1) .... 57 kts / 106 km/h
                              (C1) .... 52 kts / 96 km/h
4. After touchdown:
   Throttle .... idle
   Control stick .... full aft
   Brakes .... Gradually apply full braking—Do not lock the wheels
   Flaps .... retract

Forward Slips to a Landing

Forward slips to a landing can be performed in the Katana DA20 with or without flaps. Due to the design of the aircraft, slips have relatively little impact on the glide path of the aircraft. Forward slips to a landing should be practiced in order to meet the requirements of the Private Pilot Practical Test but have little relevance in everyday flying of the Katana DA20.

Section 5 **Katana DA20 Checklists**

Balked Landing/Go Around
1. Propeller speed control lever (A1 only) .... maximum rpm
2. Mixture (C1 only) .... full rich
3. Throttle .... full
4. Carburetor heat (A1 only) .... OFF
5. Wing flaps .... takeoff
6. Airspeed   (A1) .... 57 kts / 66 mph / 106 km/h
                     (C1) .... 52 kts / 96 km/h

After Landing and Clearing the Runway
1. Throttle .... idle
2. Mixture (C1 only) .... full rich
3. Wing flaps .... CRUISE (Up)
4. Carburetor heat (A1 only) .... OFF
5. Exterior lights .... as required
6. Electric fuel pump (A1 only) .... OFF
7. Transponder .... standby

Engine Shut-down
1. Throttle .... idle
2. Mixture (C1 only) .... idle cut-off
3. Parking brake .... set
4. ELT .... check (listen on 121.5)
5. Avionics master .... OFF
6. Electrical instruments .... OFF
7. Ignition switch .... OFF
8. Instrument panel lighting .... OFF
9. Master switch .... OFF
10. Flight control lock .... install, as required
11. Tie downs and wheel chocks .... install, as required
12. Pitot-static probe cover .... install, as required
13. Stall warning plug .... install, as required

5-15

Diamond Katana DA20: A Pilot's Guide

Important

Remember: Full reference must be made to the *Airplane Flight Manual* for the aircraft, Pilot's Operating Handbook, Flight School Syllabus, etc., for all normal and emergency procedures.

If in doubt—ask.

Section 6
Katana DA20
Loading and Performance

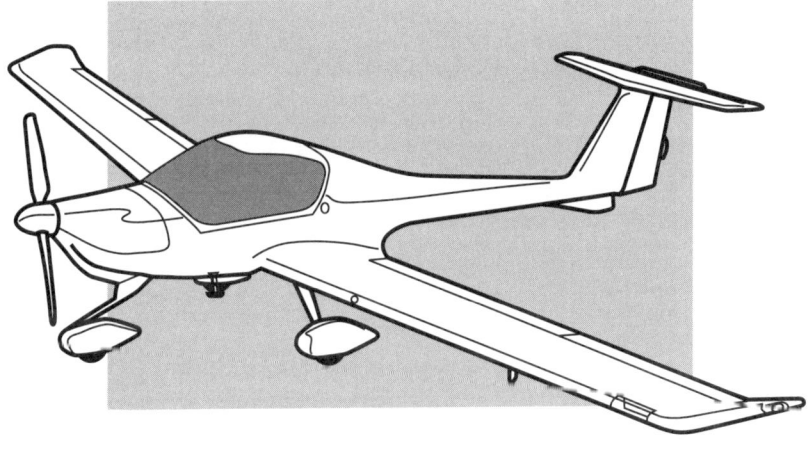

Diamond Katana DA20: A Pilot's Guide

Section 6 **Loading and Performance**

Loading

Aircraft loading can be divided into two areas, the aircraft weight and the center of gravity (CG) position.

The Katana DA20 must be loaded so that its weight is below the certified maximum takeoff weight of 1,609 lbs (730 kg) for the DA20-A1 and 1,653 lbs (750 kg) for the DA20-C1. The weight limit is set primarily as a function of the lifting capability of the aircraft, which is largely determined by the wing design and engine power of the aircraft. Operating the aircraft when it is overweight will adversely effect the handling and performance of the aircraft, such as:

- Increased takeoff speed and slower acceleration
- Increased runway length required for take off
- Reduced rate of climb
- Reduced maximum altitude capability
- Reduced range and endurance
- Reduction in maneuverability and controllability
- Increased stall speed
- Increased approach and landing speed
- Increased runway length required for landing

The aircraft must be loaded to ensure that its center of gravity is within set limits, normally defined as a forward and aft limit in inches aft of the datum. The forward limit is determined by the amount of elevator control available at landing speed. The stability and controllability of the aircraft while maneuvering determine the aft limit. Attempted flight with the CG position outside of the set limits (either forward or aft) will lead to control difficulties, and possibly loss of control of the aircraft.

When loading the aircraft it is standard practice to calculate the weight and CG position of the aircraft at the same time, commonly known as the weight and balance calculation. Before going further, it must be emphasized that the following examples are provided for illustrative purposes only. Each individual aircraft has an individual weight and balance record that is only valid for that aircraft, and is dependent upon, among other things, the equipment installed in the aircraft. If the aircraft has any major modifications, repair or new equipment installed, a new weight and balance record will be produced.

As well as setting out limits, the aircraft documents will also give arms for each item of loading. The arm is a distance from the aircraft datum to the item. The weight multiplied by its arm gives its moment. Thus, a set weight will have a greater moment the farther it is from the datum.

Diamond Katana DA20: A Pilot's Guide

Date	Entry No.		Description of part or modification	Changes of Weight						Actual Empty Weight		
				Addition (+)			Subtraction (-)					
	In	Out		Weight [lbs] ([kg])	Arm [in.] ([m])	Moment [in.lbs] ([kgm])	Weight [lbs] ([kg])	Arm [in.] ([m])	Moment [in.lbs] ([kgm])	Weight [lbs] ([kg])	Arm [in.] ([m])	Moment [in.lbs] ([kgm])

DA 20 Katana — Serial No.: _____ Registration: _____ Page No.:

Weight and Balance Report, from DA20 *Airplane Flight Manual*

The operating weight of the aircraft can be split into two categories:

EMPTY WEIGHT—the weight of the aircraft, including unusable fuel; normally this includes full oil as well. The weight and CG position of the aircraft in this condition will be noted in the weight and balance record.

USEFUL LOAD—weight of pilot, co-pilot, passengers, usable fuel and baggage. Again, the weight and balance record will give an arm for each of these loads.

Section 6 **Loading and Performance**

Mathematical Weight and Balance Calculation

With this method of calculation, the weights of each item are listed together with their arms. Addition of all the weights is the first step, to ensure that the resulting figure is within the maximum permitted. Assuming this is the case, the balance can then be calculated. For each item (except for the empty weight where the calculation is done already on the weight and balance record), the weight is multiplied by the arm, to give a moment. The *Katana DA20 Airplane Flight Manual* provides a **Weight & Balance Diagram** for pilot and passenger, fuel and baggage. Normally the arm is aft of the datum, to give a positive figure. If the arm quoted is forward of the datum, the moment will be negative (obviously the weight is not deducted from the weight calculation). All the moments are then added together to give the total moment, and this figure is then divided by the total weight. The resulting figure will be the position of the CG, which can be checked to ensure it is within the set limits.

The *Katana DA20 Airplane Flight Manual* provides a **Calculation of Loading Condition** worksheet for summarizing the weight and balance information. The weight and CG position can be plotted on the **Permissible Center of Gravity Range and Permissible Flight-Weight-Moment** graph in the *Airplane Flight Manual*. If the plotted position is within the "envelope," the weight and CG position are within limits.

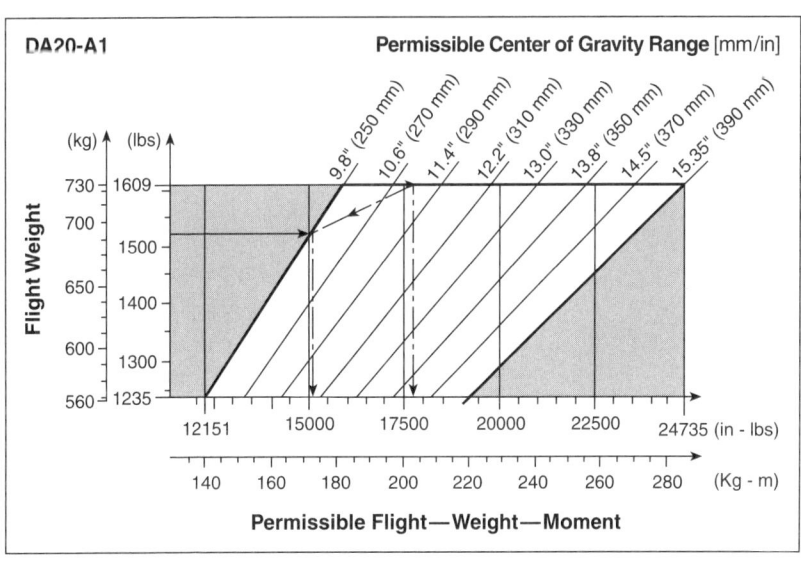

Diamond Katana DA20: A Pilot's Guide

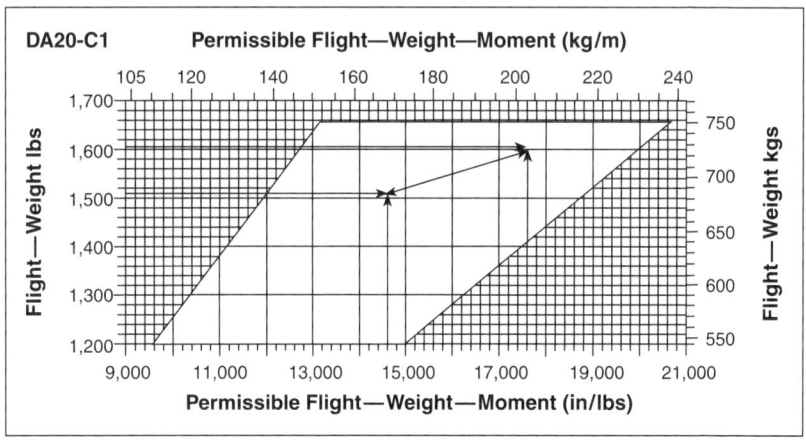

A Word Of Warning. As well as the safety aspect, operating the aircraft outside its weight and balance envelope has far-reaching legal and financial implications. Almost the first thing an accident investigator will check after an accident is the loading of the aircraft. If the loading is outside the limits, the pilot is violating the Federal Aviation Regulations. In addition, both the aircraft insurance company and your personal insurance company will be unsympathetic when they know that the conditions of the Airworthiness Certificate (i.e., the flight manual limitations) were not complied with. As the pilot-in-command the responsibility is yours alone. The fact that the Katana has two seats does not necessarily mean the aircraft can be flown with two football players, baggage, and a full fuel load.

Performance

The *Airplane Flight Manual* has a section of tables and diagrams to allow the pilot to calculate the expected performance of the aircraft for different phases of the flight such as takeoff, climb and landing. It is important to understand that the manufacturer's performance calculations were arrived at by exactly replicating recommended techniques, and the test aircraft were new and operated by experienced test pilots; moreover, they were flown under favorable conditions. To make allowances for a less-than-new aircraft being flown by an average general aviation pilot, it is wise to "factor" any results you get.

As with loading calculations, the pilot must use the tables and diagrams for the aircraft being used. The graphs and diagrams used in this section are for illustration purposes only, and are not intended for operational use.

Section 6 **Loading and Performance**

Takeoff Performance

The takeoff performance can be divided into two sections: The Takeoff Run (or Takeoff Ground Roll), the distance taken for the aircraft to become airborne, and; The Takeoff Distance, that is the total distance required for the aircraft to become airborne *and* clear a 50' barrier.

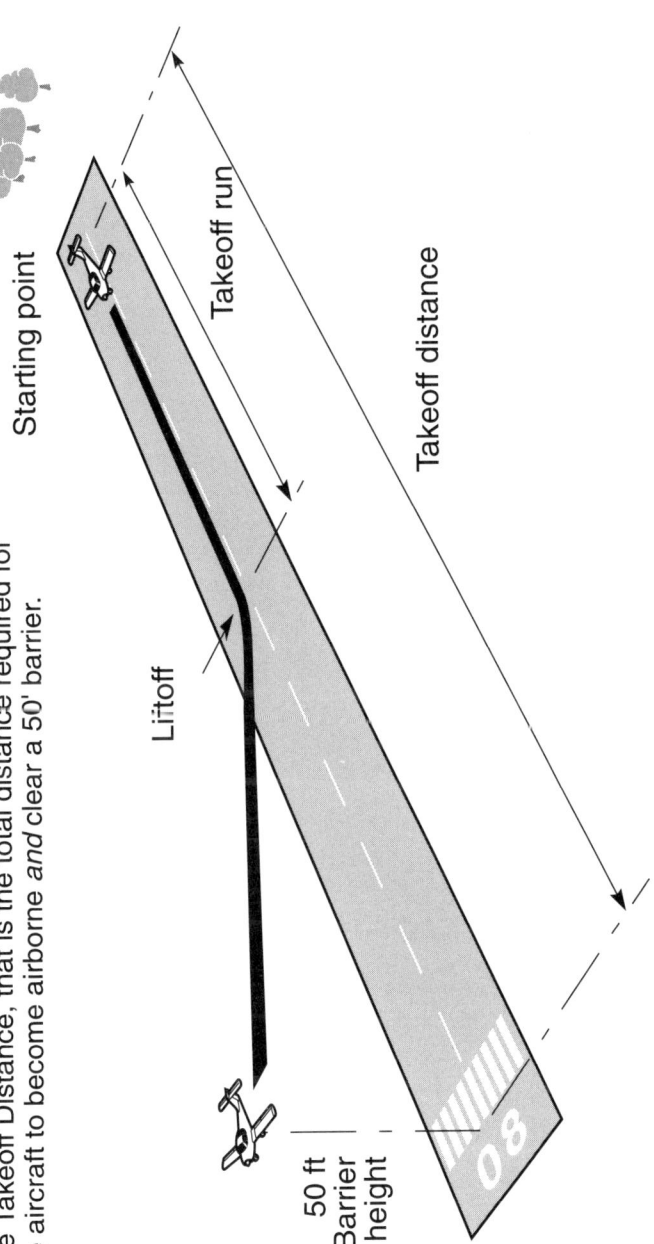

6-7

Diamond Katana DA20: A Pilot's Guide

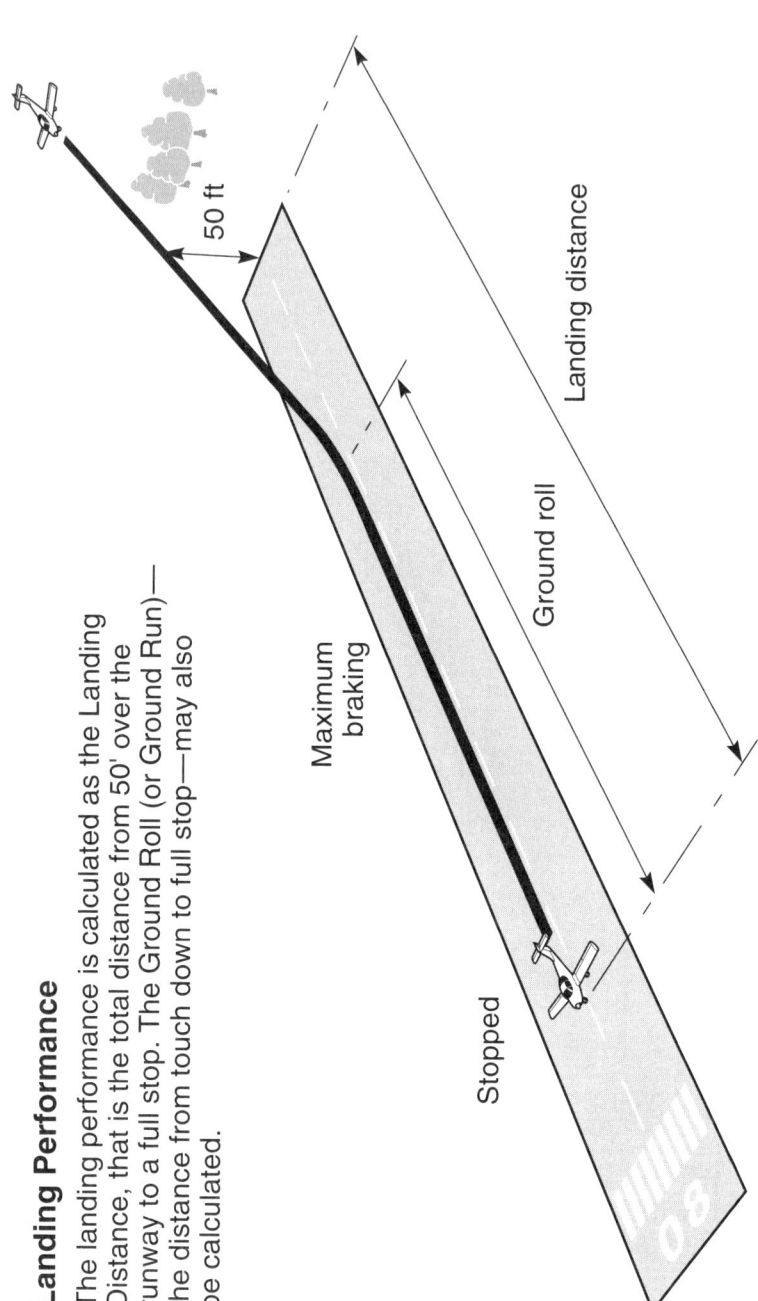

Landing Performance

The landing performance is calculated as the Landing Distance, that is the total distance from 50' over the runway to a full stop. The Ground Roll (or Ground Run)—the distance from touch down to full stop—may also be calculated.

Section 6 **Loading and Performance**

The manufacturer's performance data were all corrected to standard atmospheric conditions (59°F/15°C and 29.92 in. Hg/1013.25 mb) at sea level. Section 7 of this book provides conversion factors and recommended factors for variations not necessarily covered by flight manual graphs.

Takeoff and Landing Performance Graphs

The takeoff distance and landing distance graphs in the flight manual make several assumptions: paved dry, level runway, and use of flight manual technique. (*See* landing distance graph below, and takeoff distance graph on the next page.)

The graphs use the term "Pressure Altitude." This is the altitude of the runway assuming a standard barometric pressure setting of 29.92 inches Hg. On a day when the pressure is other than 29.92, you will have to adjust the actual altitude to get the pressure altitude. For instance, on a day with the pressure above 29.92, the pressure altitude will be less than the actual, and vice versa. To do this conversion, simply adjust the actual altitude by 1,000 feet for each inch Hg. above or below 29.92 (10 feet for each .01 inch).

Landing and Rolling Distances for Heights Above MSL, DA20-A1

Height above MSL	ft.	0	1000	2000	3000	4000	5000
	(m)	(0)	(305)	(610)	(915)	(1220)	(1524)
Landing Distance	ft.	1490	1550	1609	1669	1728	1788
	(m)	(454)	(472)	(491)	(509)	(527)	(545)
Landing Roll Distance	ft.	748	770	793	817	842	868
	(m)	(228)	(235)	(242)	(249)	(257)	(265)

Landing and Rolling Distances for Heights Above MSL, DA20-C1

Height above MSL	ft.	0	1000	2000	3000	4000	5000	6000	7000
	(m)	(0)	(305)	(610)	(915)	(1220)	(1524)	(1829)	(2134)
Landing Distance	ft.	1280	1305	1332	1360	1388	1418	1449	1481
	(m)	(390)	(398)	(406)	(414)	(423)	(432)	(442)	(451)
Landing Roll Distance	ft.	581	598	616	635	654	674	695	716
	(m)	(177)	(182)	(188)	(193)	(199)	(205)	(212)	(225)

Note: Poor maintenance condition of the airplane, deviation from the given procedures as well as unfavorable outside conditions (i.e. high temperature, rain, unfavorable wind conditions, slippery runway) could increase the landing distance considerably.

The headwind or tailwind component is calculated from the wind speed and the angle to the runway (i.e., a 10-knot wind directly down the runway gives a headwind component of 10 knots; a 10-knot wind at 90° to the runway gives a headwind component of 0). There is a graph in Section 7 for calculating the head/tailwind component and the crosswind component.

The takeoff distance and landing distance graphs will state the technique used to obtain the figures. Remember to get the same performance out of the airplane as the performance charts and tables indicate, professional pilot techniques and identical field conditions must exist.

Enroute Performance Graphs

Data is also provided in the flight manual for calculating the enroute performance, such as climb performance, cruising speed, and maximum flight duration. These graphs use the same technique as for the takeoff and landing graphs. Again, remember that to experience the projected results from the aircraft handbook in actual flight conditions, the same conditions and technique must be used. (*See* graphs on Pages 6-11 through 6-18.)

Runway Dimensions

Once you have calculated the distance the aircraft requires for takeoff or landing, the runway dimensions must be checked to ensure that the aircraft can be safely operated on the runway in question. The figure given in the **Airport/Facility Directory** or airfield guide can be defined in a number of ways.

Takeoff Run Available (TORA)
The TORA is the length of the runway available for the takeoff ground run of the aircraft. This is usually the physical length of the runway.

Accelerate/Stop Distance (A/SD)
The A/SD is the length of the TORA plus the length of any stopway. A stopway is the area at the end of the TORA prepared for an aircraft to stop on in the event of an abandoned takeoff.

Takeoff Distance Available (TODA)
The TODA is the TORA plus the length of any clearway. A clearway is an area over which an aircraft may make its initial climb (to 50 feet in this instance).

Landing Distance Available (LDA)
The LDA is the length of the runway available for the ground run of an aircraft landing. In all cases, the landing distance required should never be greater than the landing distance available.

Section 6 **Loading and Performance**

Takeoff Distance, DA20-A1

(From DA20-A1 *Aircraft Flight Manual*)

Conditions:
- maximum takeoff power
- lift-off speed 57 KIAS and speed for climb over obstacle 60 KIAS
- level runway, paved
- wing flaps in takeoff position (T/O)

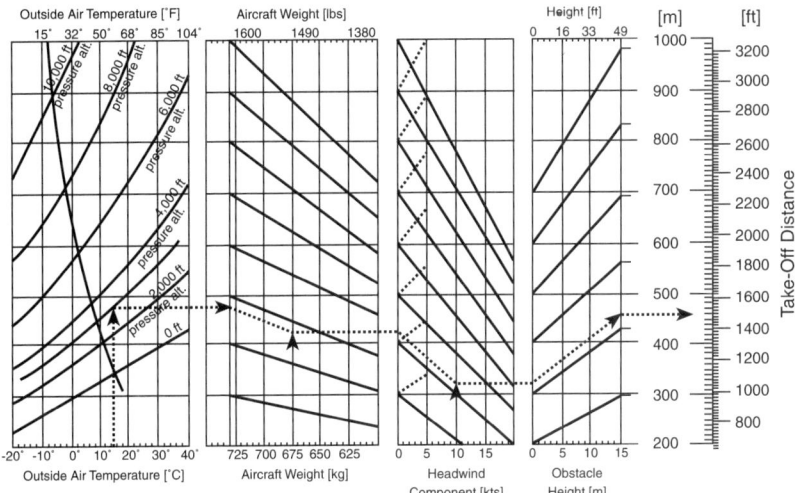

Example:	Pressure altitude	3,000 ft
	Outside temperature	15°C (59°F)
	Weight	1,488 lbs (675 kg)
	Wind	10 kts
Result:	Takeoff roll distance	1,080 ft (330 m)
	Takeoff distance to clear a 15 m (50 ft) obstacle	1,540 ft (470 m)

Note: Poor maintenance condition of the airplane, deviation from the given procedures as well as unfavorable outside conditions (i.e. high temperature, rain, unfavorable wind conditions, including crosswind) could increase the takeoff distance considerably. For takeoff from dry, short-cut grass-covered runways compared to paved runways, a 25% increase in takeoff roll distance must be taken into account.

On soft grass-covered runways with grass deeper than 4 in, (10 cm) the takeoff roll distance might be increased by as much as 40%.

Diamond Katana DA20: A Pilot's Guide

Takeoff Distance, DA20-C1
Hoffmann Propeller HO-141HM-175-157

(From DA20-C1 *Airplane Flight Manual*)

Conditions:
- maximum takeoff power
- lift-off speed 52 KIAS and speed for climb over obstacle 58 KIAS
- level runway, paved
- wing flaps in takeoff position (T/O)

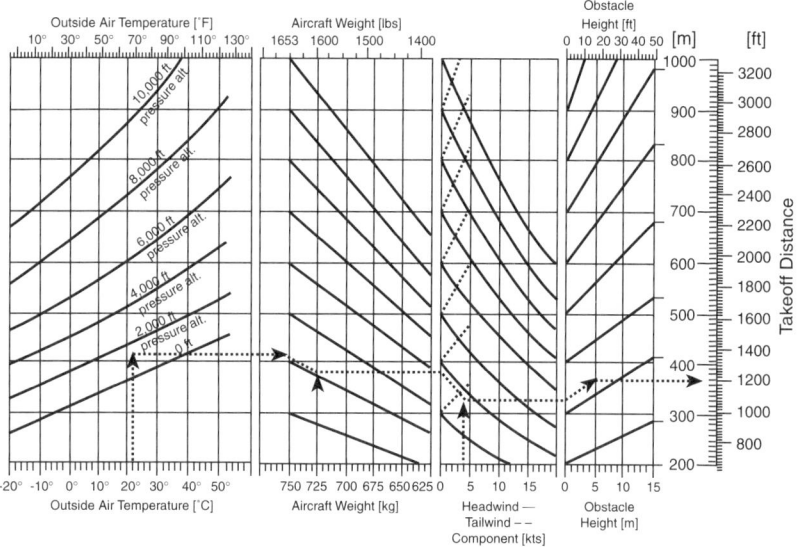

Example: Pressure altitude .... 1,000 ft
            Outside temperature .... 72°F (22°C)
            Weight .... 1,600 lbs (725 kg)
            Wind .... 4 kt headwind

Result: Takeoff roll distance to clear a 16 ft (5 m) obstacle .... 1,214 ft (370 m)

Note: Poor maintenance condition of the airplane, deviation from the given procedures as well as unfavorable conditions (i.e., high temperature, rain, unfavorable wind conditions, including crosswind) could increase the takeoff distance considerably.

Section 6 **Loading and Performance**

Takeoff Distance, DA20-C1
Sensenich Propeller W69EK-63

(From DA20-A1 *Airplane Flight Manual*)

Conditions:
- maximum takeoff power
- lift-off speed 52 KIAS and speed for climb over obstacle 58 KIAS
- level runway, paved
- wing flaps in takeoff position (T/O)

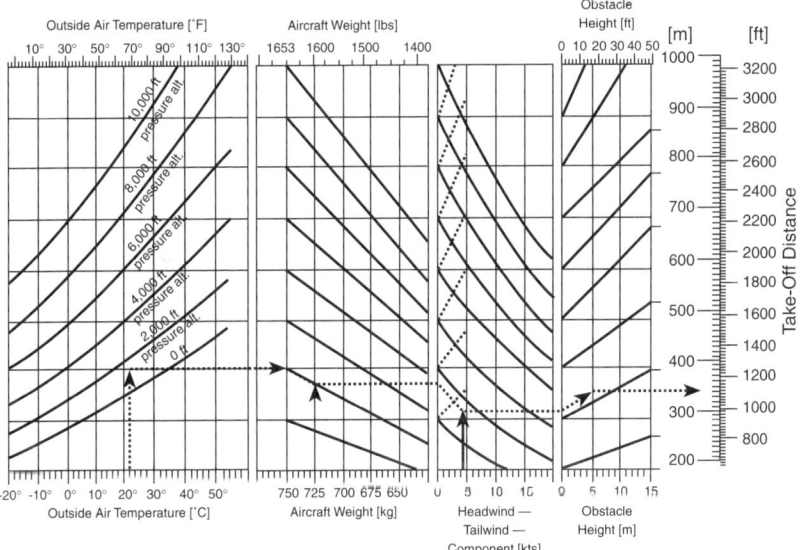

Example: Pressure altitude... 1,000 ft
Outside temperature.. 22°C (72°F)
Weight.. 1,600 lbs (725 kg)
Wind... 4 kts

Result: Takeoff distance to clear a 5 m (16 ft) obstacle..... 1,148 ft (350 m)

Note: Poor maintenance condition of the airplane, deviation from the given procedures as well as unfavorable outside conditions (i.e. high temperature, rain, unfavorable wind conditions, including crosswind) can increase the takeoff distance considerably.

Diamond Katana DA20: A Pilot's Guide

Climb Performance/Takeoff, DA20-C1 Sensenich Propeller W69EK-63

(From DA20-C1 *Airplane Flight Manual*)

Best rate-of-climb speed with wing flaps T/O.......................68 KIAS

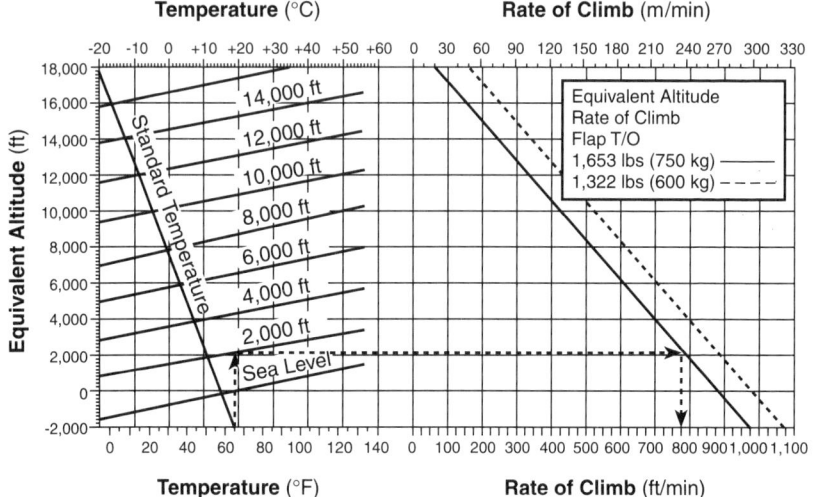

Example:	Pressure Altitude....................2,000 ft
	OAT..65°F
	Weight.....................................1,653 lbs
Result:	Climb performance.................695 ft/min

Note: At the time of this printing, the manufacturer has not yet provided a "climb performance/takeoff" table for the DA20-A1, or the DA20-C1 (Hoffmann Propeller).

Section 6 **Loading and Performance**

Climb Performance/Cruising Altitudes, DA20-A1

Max. cruising altitude (in standard conditions)..............................13,120 ft (4,000 m)
Best rate-of-climb speed with wing flaps in
takeoff position (T/O)..65 kts/75 mph/120 km/h

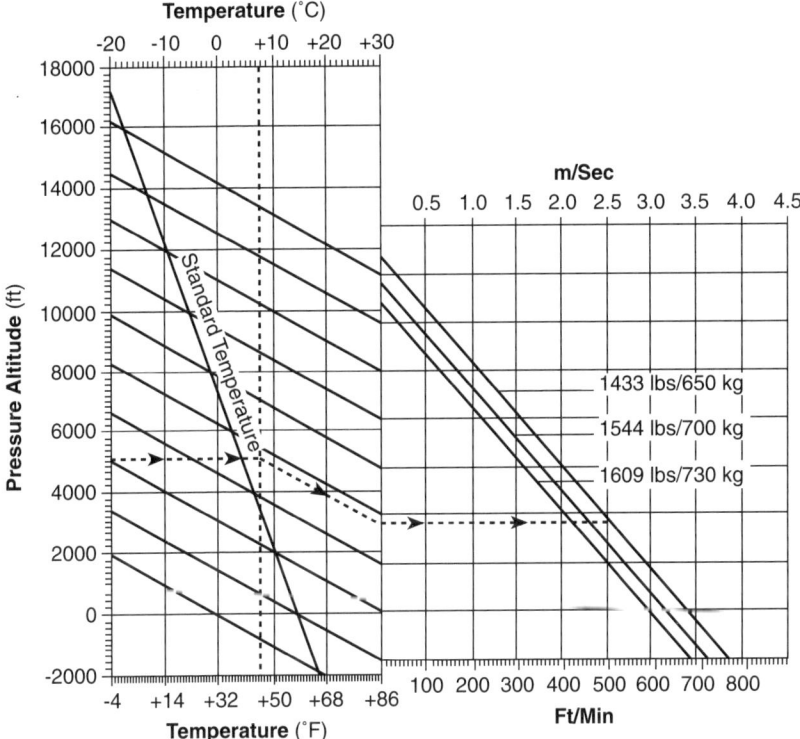

Example: Pressure Altitude....................5,000 ft (1524 m)
OAT......................................46°F (8°C)
Weight..................................1,477 lbs (670 kg)
Result: Climb performance................490 ft/min (2.5 m/s)

Caution: In case of operation without wheel fairings, the climb performance is reduced by approximately 3%.

6-15

Diamond Katana DA20: A Pilot's Guide

Climb Performance/Cruising Altitudes, DA20-C1
Hoffmann Propeller HO-141HM-175-157

Max. cruising altitude (in standard conditions)..............................13,120 ft (4,000 m)
Best rate-of-climb speed with wing flaps CRUISE........................75 KIAS

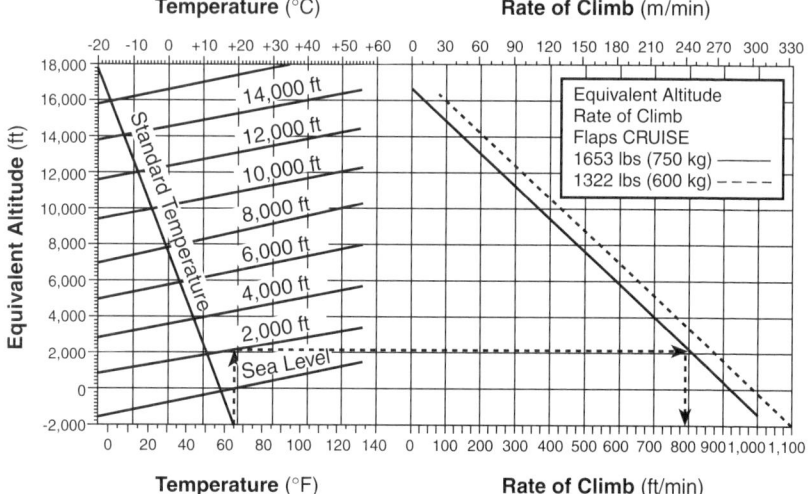

Example:	Pressure Altitude....................2,000 ft
	OAT..65°F
	Weight......................................1,653 lbs
Result:	Climb performance.................785 ft/min

6-16

Section 6 **Loading and Performance**

Climb Performance/Cruising Altitudes, DA20-C1 Sensenich Propeller W69EK-63

(From DA20-C1 *Airplane Flight Manual*)

Max. cruising altitude (in standard conditions)..............................13,120 ft (4,000 m)
Best rate-of-climb speed with wing flaps CRUISE.......................75 KIAS

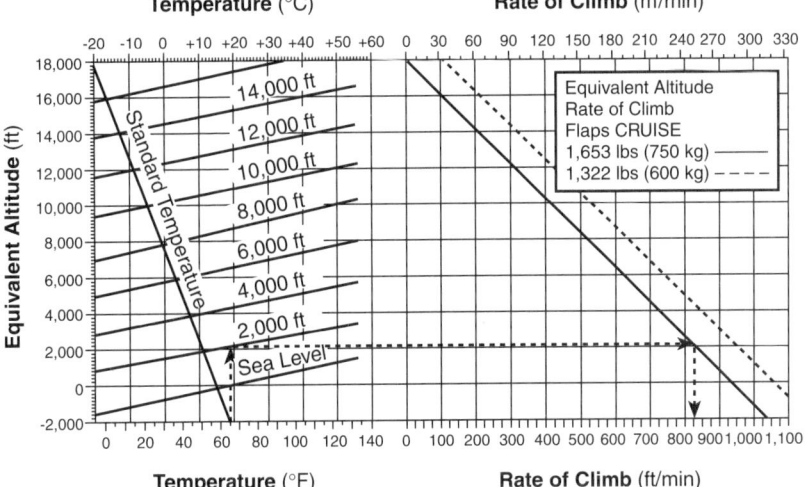

Example:	Pressure Altitude.....................2,000 ft
	OAT...65°F
	Weight.....................................1,653 lbs
Result:	Climb performance.................830 ft/min

6-17

Diamond Katana DA20: A Pilot's Guide

Cruising Speed (True Airspeed), DA20-A1

Diagram for true airspeed (TAS) calculation at selected power level.

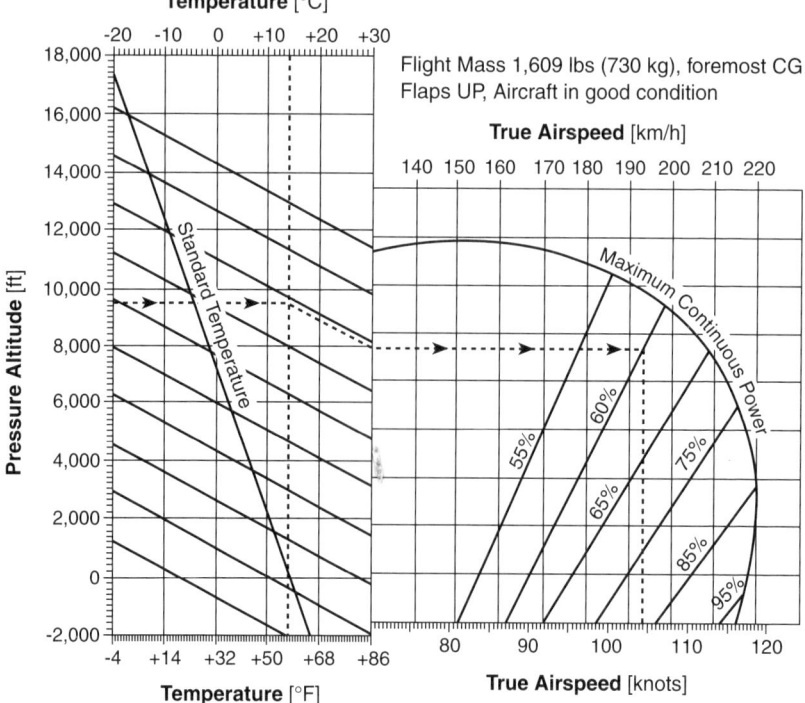

Example:	Pressure altitude:	9,500 ft
	Temperature:	57°F (14°C)
	Power setting:	60%
Result:	True airspeed (TAS):	104.2 kts (193 km/h)

Caution: In case of operation without wheel fairings, the maximum cruising speed is reduced by approximately 5%.

Section 6 **Loading and Performance**

Cruising Speed (True Airspeed), DA20-C1
Hoffmann Propeller HO-141-HM-175-157

(From DA20-C1 *Airplane Flight Manual*)

Diagram for true airspeed (TAS) calculation at selected power level.

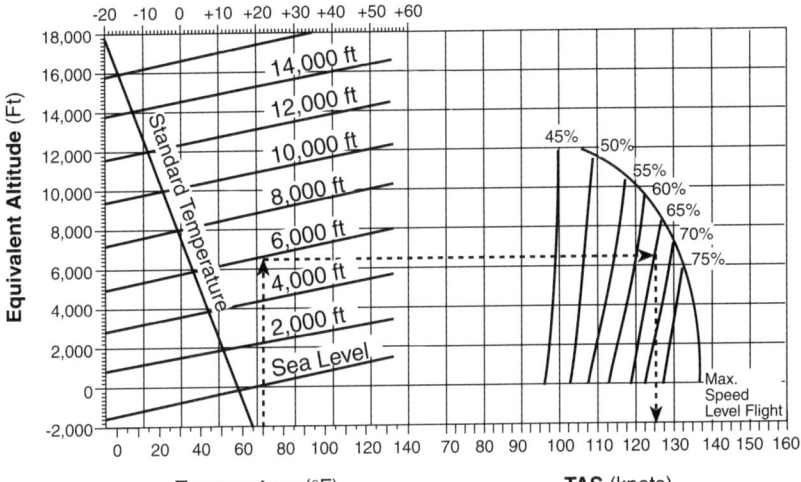

Example:	Pressure altitude:	6,000 ft
	Temperature:	70°F
	Power setting:	65%
Result:	True airspeed (TAS):	124 kts

6-19

Diamond Katana DA20: A Pilot's Guide

Cruising Speed (True Airspeed), DA20-C1 Sensenich Propeller W69EK-63

(From DA20-C1 *Airplane Flight Manual*)
Diagram for true airspeed (TAS) calculation at selected power level.

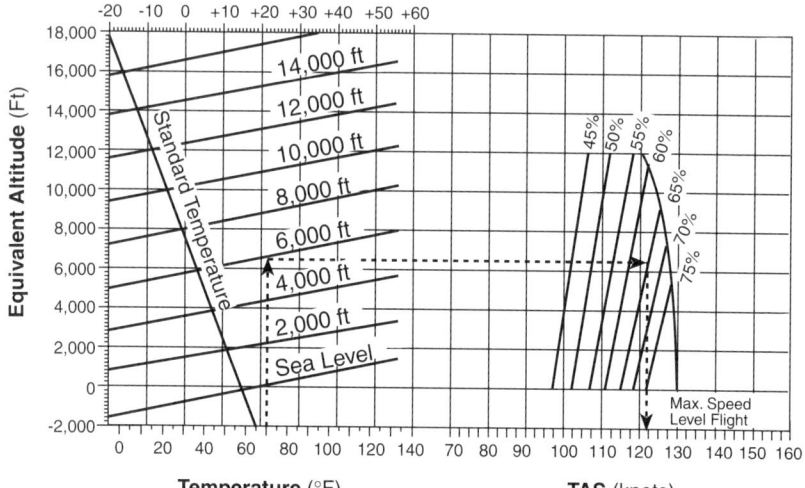

Example:	Pressure altitude:	6,000 ft
	Temperature:	70°F
	Power setting:	65%
Result:	True airspeed (TAS):	121 kts

Section 6 **Loading and Performance**

Maximum Flight Duration (DA20-A1 only)*

Diagram for calculation of the maximum flight duration depending on fuel availability.

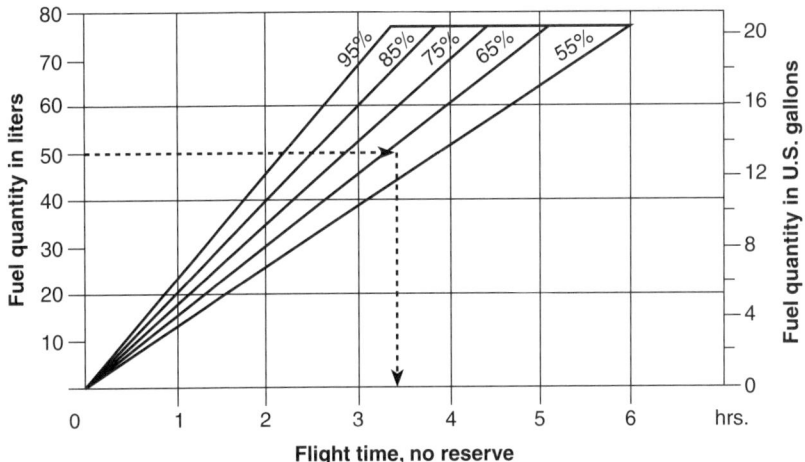

Flight time, no reserve

Example:	Fuel capacity:	13.2 U.S. gal (50 liters)
	Power Setting:	65%
Result:	Possible flight time **without** reserve:	3:28 h:min
	Possible flight time **with** reserve of 45 mins:	2:43 h:min

***Note:** At the time of this printing, the manufacturer has not yet provided a flight duration table for the DA20-C1 model.

Section 7
Conversions

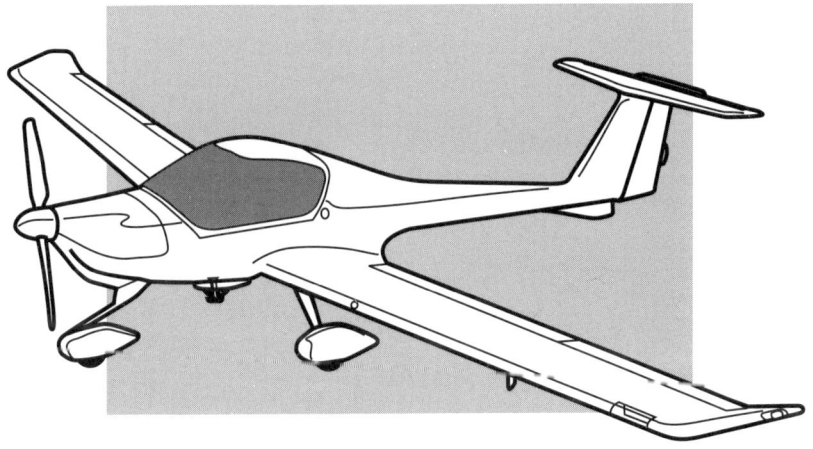

Diamond Katana DA20: A Pilot's Guide

Section 7 **Conversions**

Takeoff Distance Factors

The following factors will allow the pilot to make allowance for variations that may affect takeoff performance. Although some of these factors are covered in the DA20 performance tables, the table is produced in its entirety for completeness:

VARIATION	INCREASE IN TAKEOFF DISTANCE (to 50')	FACTOR
10% increase in aircraft weight	20%	1.2
Increase of 1,000' in runway altitude	10%	1.1
Increase in temperature of 10°C	10%	1.1
Dry Grass		
—Short (under 5 inches)	20%	1.2
—Long (5 – 10 inches)	25%	1.25
Wet Grass		
—Short	25%	1.25
—Long	30%	1.3
2% uphill slope	10%	1.1
Tailwind component of 10% of lift-off speed	20%	1.2
Soft ground or snow *	at least 25%	at least 1.25

* snow and other runway contamination are covered on page 7-5.

7-3

Diamond Katana DA20: A Pilot's Guide

Landing Distance Factors

The following factors will allow the pilot to make allowance for variations that may affect landing performance. Although some of these factors are covered in the DA20 performance tables, the table is produced in its entirety for completeness:

VARIATION	INCREASE IN LANDING DISTANCE (from 50')	FACTOR
10% increase in aircraft weight	10%	1.1
Increase of 1,000' in runway altitude	5%	1.05
Increase in temperature of 10°C	5%	1.05
Dry Grass		
—Short (under 5 inches)	20%	1.2
—Long (5 – 10 inches)	30%	1.3
Wet Grass		
—Short	30%	1.30
—Long	40%	1.40
2% downhill slope	10%	1.1
Tailwind component of 10% of landing speed	20%	1.2
snow *	at least 25%	at least 1.25

* snow and other runway contamination are covered on page 7-5.

Runway Contamination

A runway can be contaminated by water, snow or slush. If operation on such a runway cannot be avoided, additional allowance must be made for the problems such contamination may cause—i.e., additional drag, reduced braking performance (possible hydroplaning), and directional control problems.

It is generally recommended that takeoff should not be attempted if dry snow covers the runway to a depth of more than 2", or if water, slush or wet snow covers the runway to more than 1/2". In addition a tailwind, or crosswind component exceeding 10 knots, should not be accepted when operating on a slippery runway.

For takeoff distance required calculations, the other known conditions should be factored, and the accelerate/stop distance available on the runway should be at least 2.0 x the takeoff distance required (for a paved runway) or at least 2.66 x the takeoff distance required (for a grass runway).

Any water or slush can have a very adverse effect on landing performance, and the danger of hydroplaning (with negligible wheel braking and loss of directional control) is very real.

Diamond Katana DA20: A Pilot's Guide

Use of the Wind Component Graph

This graph can be used to find the head/tail wind component and the crosswind component, given a particular wind velocity and runway direction.

EXAMPLE:

Runway 27

Surface wind 240°/15 knots

The angle between the runway direction (270°) and wind direction (240°) is 30°. Now on the graph locate a point on the 30° line, where it crosses the 15 knot arc. From this point take a horizontal line to give the headwind component (13 knots) and a vertical line to give the crosswind component (8 knots).

On the main graph overleaf the shaded area represents the maximum demonstrated crosswind component for this aircraft. If the wind point is within this shaded area, the maximum demonstrated crosswind component for this aircraft has been exceeded.

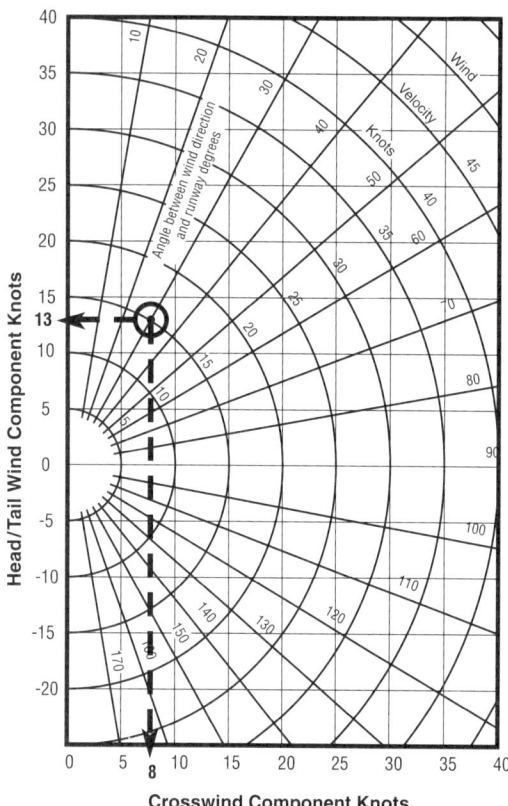

Note:
Runway direction will be degrees magnetic. Check the wind direction given is also in degrees magnetic.

Section 7 **Conversions**

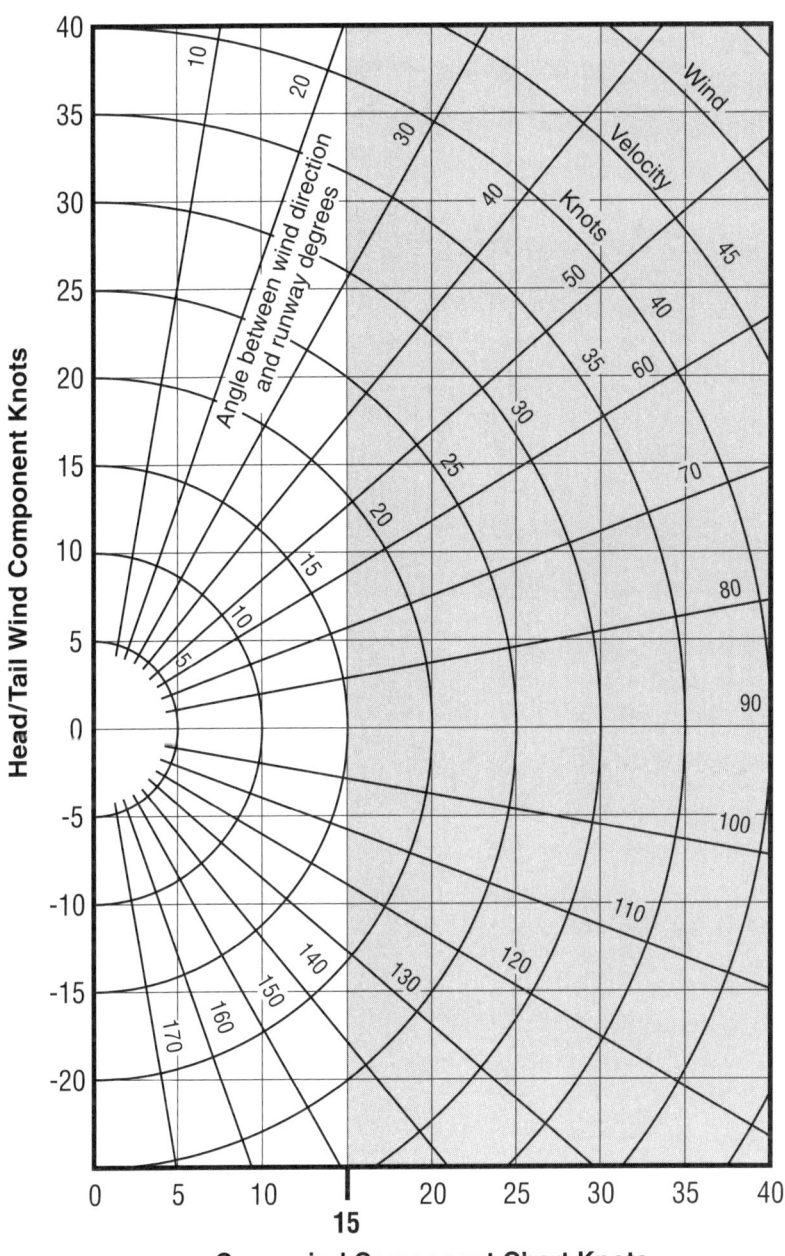

Diamond Katana DA20: A Pilot's Guide

Temperature

°C	°F
50	120
	110
40	100
	90
30	80
20	70
	60
10	50
	40
0	30
	20
−10	10
	0
−20	−10
	−20
−30	−30
−40	−40
	−50
−50	−60
	−70
−60	

Section 7 **Conversions**

Distance – Meters/Feet

Meters	Feet	Feet	Meters
1	3.28	1	0.30
2	6.56	2	0.61
3	9.84	3	0.91
4	13.12	4	1.22
5	16.40	5	1.52
6	19.69	6	1.83
7	22.97	7	2.13
8	26.25	8	2.44
9	29.53	9	2.74
10	32.81	10	3.05
20	65.62	20	6.10
30	98.43	30	9.14
40	131.23	40	12.19
50	164.04	50	15.24
60	196.85	60	18.29
70	229.66	70	21.34
80	262.47	80	24.38
90	295.28	90	27.43
100	328.08	100	30.48
200	656.16	200	60.96
300	984.25	300	91.44
400	1,312.34	400	121.92
500	1,640.42	500	152.40
600	1,968.50	600	182.88
700	2,296.59	700	213.36
800	2,624.67	800	243.84
900	2,952.76	900	274.32
1,000	3,280.84	1,000	304.80
2,000	6,561.70	2,000	609.60
3,000	9,842.50	3,000	914.40
4,000	13,123.40	4,000	1,219.20
5,000	16,404.20	5,000	1,524.00
6,000	19,685.00	6,000	1,828.80
7,000	22,965.90	7,000	2,133.60
8,000	26,246.70	8,000	2,438.40
9,000	29,527.60	9,000	2,743.20
10,000	32,808.40	10,000	3,048.00

Conversion Factors:

Centimeters to Inches x .3937
Inches to Centimeters x 2.54

Meters to Feet x 3.28084
Feet to Meters x 0.3048

Diamond Katana DA20: A Pilot's Guide

Distance – Nautical Miles / Statute Miles

NM	SM
1	1.15
2	2.30
3	3.45
4	4.60
5	5.75
6	6.90
7	8.06
8	9.21
9	10.36
10	11.51
20	23.02
30	34.52
40	46.03
50	57.54
60	69.05
70	80.55
80	92.06
90	103.57
100	115.1
200	230.2
300	345.2
400	460.3
500	575.4
600	690.5
700	805.6
800	920.6
900	1035.7

SM	NM
1	.87
2	1.74
3	2.61
4	3.48
5	4.34
6	5.21
7	6.08
8	6.95
9	7.82
10	8.69
20	17.38
30	26.07
40	34.76
50	43.45
60	52.14
70	60.83
80	69.52
90	78.21
100	86.9
200	173.8
300	260.7
400	347.6
500	434.5
600	521.4
700	608.3
800	695.2
900	782.1

Conversion Factors:
Statute Miles to Nautical Miles x 0.868976
Nautical Miles to Statute Miles x 1.15078

Section 7 **Conversions**

Volume (Fluid)

Liters	U.S. Gal.
1	0.26
2	0.53
3	0.79
4	1.06
5	1.32
6	1.59
7	1.85
8	2.11
9	2.38
10	2.64
20	5.28
30	7.93
40	10.57
50	13.21
60	15.85
70	18.49
80	21.14
90	23.78
100	26.42
200	52.84
300	79.26
400	105.68
500	132.10
600	158.52
700	184.94
800	211.36
900	237.78
1000	264.20

U.S. Gal.	Liters
1	3.79
2	7.57
3	11.36
4	15.14
5	18.93
6	22.71
7	26.50
8	30.28
9	34.07
10	37.85
20	75.71
30	113.56
40	151.41
50	189.27
60	227.12
70	264.97
80	302.82
90	340.68
100	378.54

Conversion Factors:
U.S. Gallons to Liters x 3.78541
Liters to U.S. Gallons x 0.264179

Diamond Katana DA20: A Pilot's Guide

DA20—Index

A

accelerate/stop distance (A/SD) 6-10
ailerons 1-8
airframe 1-6
airspeed limitations 2-6, 2-7
alternator 1-26, 1-27
ammeter 1-28
automotive gasoline 2-10
AVGAS 2-10, 2-12
avionics 1-32
avionics power switch 1-29

B

baggage compartment 1-37
baggage net 1-37
battery 1-27
Bombardier Rotax 912 F3 engine 1-11–1-14, 1-19
brake check 3-7
brakes 1-10

C

cabin air 1-35
cabin heating 1-35
canopy 1-37
canopy warning light 1-37
carburetor 1-13, 1-24–1-25
carburetor heat control, use of 4-16
carburetor icing 4-13
carburetor icing check 4-16
center of gravity 6-3
checklists
    accelerated/fast descent 5-13
    after landing and clearing the runway 5-15
    approaching aircraft 5-3
    balked landing/go around 5-15
    before starting engine 5-7
    before takeoff 5-9
    before taxiing 5-9
    climb 5-12
    cruise 5-13
    descent 5-13

Continued

Diamond Katana DA20: A Pilot's Guide

engine shut-down 5-15
landing approach 5-13
normal takeoff 5-11
short-field landing 5-14
short-field takeoff 5-12
soft-field landing 5-14
soft-field takeoff 5-12
starting engine 5-8
takeoff 5-11
taxiing 5-9
circuit breakers 1-26, 1-29
climb performance/cruising altitudes
    DA20-A1 6-15
    DA20-C1 6-16
climbing 3-10
cockpit vents 1-35
control lock 3-16
coolant types 2-10
cowling 1-11
crosswind component chart 7-6, 7-7
crosswind, maximum demonstrated 2-6, 2-7
cruise flight 3-10
cruise performance 6-10

D

descent 3-14
dimensions
    DA20-A1 2-3
    DA20-C1 2-4
distance conversion – meters/feet 7-9
drainage points 1-23

E

electrical system 1-26
empty weight 6-4
engine
    Continental IO-240-B 3-6
    DA20-A1 1-11
    DA20-C1 1-14
    handling 3-10
    limitations—DA20-C1, Continental IO-240-B 2-11
    Rotax 912 3-4
    starting, DA20-A1 3-4
    starting, DA20-C1 3-6

Index

F
flap control 1-8
flaps 1-8
flight controls 1-8
flight duration 6-20
flight manual 5-3
forward slips to a landing 5-14
fuel
    consumption 4-17
    exhaustion 3-10
    filter 1-21
    grades 2-10, 2-12
    icing 4-13
    quantity sensor 1-23
    selector 3-11
    system 1-21
    system limitations 2-10
    system venting 1-21
    tanks, capacity and limitations 1-21, 2-10, 2-12
fueling 1-21
fuselage 1-6

G
ground handling 3-3
ground roll/ground run 6-7
gyroscopic instruments 1-33

H
harnesses 1-36
heating system 1-35
horizontal stabilizer 1-7

I
icing 4-13
ignition system 1-18, 3-8
impact icing 4-13

L
landing
    distance available (LDA) .. 6-10
    distance factors 6-10, 7-4
    gear 1-10
    light 1-31

I-3

Diamond Katana DA20: A Pilot's Guide

    performance 6-8
    performance table 6-9
    procedure 3-15
lighting system 1-31
loading 6-3
loading graph 6-5

M

main wheel pressures 2-11–2-12
main wheels 1-10
master switch 1-29
model numbers and names 1-4

N

nose wheel 1-10
nose wheel pressures 2-11–2-12

O

oil
    grades 1-13, 2-10, 2-12
    high temperature 2-9, 2-11
    low pressure 3-5
    low temperature 2-9, 2-11
    pressure gauge 1-19
    sump 1-19
    system 1-19
    system limitations and capacity 1-19, 2-9–2-11
    temperature gauge 1-19
overvoltage sensor 1-29
overweight limitations 6-3
overweight loading 6-6

P

parking and tie down 3-16
parking brake 1-10
performance 6-6–6-10
performance limitations 2-8
pitot heat 1-34
pitot tube 1-34
pitot-static system 1-34–1-35
power checks 3-8
pressure altitude 6-9
production 1-3–1-4
propeller 1-16

Index

R
rudder 1-8
rudder pedal distance 3-4
runway contamination 7-5
runway dimensions 6-10

S
seats 1-36
side slipping 3-14
spark plugs 1-11
spins 3-13–3-14
stall warning 1-30, 3-12
stalls 3-11
starter 1-20
static vent port 1-34
strobe 1-31

T
takeoff
   distance available (TODA) 6-10
   distance, DA20-A1 6-11
   distance, DA20-C1 6-12
   distance factors 7-3
   general 3-9
   performance 6-7
   performance table 6-11
   run 6-7
   run available (TORA) 6-10
   weight 6-3
taxiing 3-7
Teledyne Continental IO-240-B engine 1-14
temperature conversion 7-8
throttle 1-16
throttle icing 4-14
tie down 3-16
tie-down points 3-16
tire pressures 1-10, 2-11–2-12
tires 1-10
toe brakes 1-10
towbar 3-3
turbulent conditions 3-10

U
useful load 6-4

Diamond Katana DA20: A Pilot's Guide

V

"V" speeds code 2-5
ventilation system 1-35
visual inspection 5-3
voltage regulator 1-27
voltmeter 1-28
volume conversion 7-11

W

weight and balance calculation 6-5
weight and balance record 6-3, 6-4
wind component graph 7-6
windows 1-37
wing 1-6

Notes

Notes

Notes

Notes

Notes

Notes